Build • Connect • Understand

2 Digit Addition & Subtraction

A Mathematics Unit Based on Inquiry, Problem Solving, and Student Discourse

Lisa Ann de Garcia, MEd

2 Digit Addition & Subtraction: A mathematics unit based on inquiry, problem solving, and student discourse

All rights reserved.
No part of this manual may be reproduced or transmitted by any form or any means, electronic or mechanical, including photocopy, recording or any information storage and retrieval without the written consent of the copyright holder.

Copyright 2015 Lisa Ann de Garcia
www.developmentalmathassessment.com

ISBN 978-0-9860917-6-6

Table of Contents

Introducing the unit | 1
Build, Connect, Understand | 2
The Lesson Map | 5
Setting up an Environment of Discourse | 9
Using Problem Solving as a Vehicle for Understanding Computation | 13
Unit Overview | 15
Daily Snapshot | 16

Lessons

   Develop:
      Groups & Leftovers | 19
      Representing a Number | 22
      Join Result Unknown no regrouping / no recording | 24
      Join Result Unknown no regrouping / initial recording with pictures | 27
      Separate Result Unknown no regrouping / recording with pictures | 30
      Separate Result Unknown no regrouping / recording with pictures (2) | 33
      Join Result Unknown regrouping recording with pictures / initial recording with numbers | 36
      Join Result Unknown regrouping recording with pictures / initial recording with numbers (2) | 39
      Join Result Unknown regrouping / recording with pictures and numbers | 42
      Join Change Unknown recording with pictures and numbers | 45
      Separate Change Unknown | 48
      Separate Change Unknown another look | 51
      Separate Start Unknown | 54

   Solidify:
      Separate Start Unknown try on another student's thinking | 57
      Continue to Try on Thinking | 60
      Continue to Try on Thinking (2) | 64
      Using Graphs to generate questions | 69
      Using Graphs to compare data | 72
      Using Graphs to Compare trying on another student's thinking | 76
      More Comparing with graphs | 79
      A look at Renaming | 83
      Trying the Renaming Strategy | 88
      Further Exploration of the renaming strategy | 91
      More Data to Analyze | 95

   Practice:
      Practicing Subtraction | 100
      Practicing Subtraction by analyzing a mistake | 106
      Practicing Subtraction analyzing another mistake | 111
      Practicing Subtraction Procedures | 116

References | 123

Introducing the Unit

This is the first *Build, Connect, Understand* book in what I hope to be a robust series of math concepts across the grades. This is a 30 day unit on two-digit addition and subtraction that was conducted with a second grade class. This unit is centered around problem solving, building understanding, communicating thinking and ideas, and making connections so students are able to truly understand the algorithms they use when adding and subtracting.

The students of this particular classroom had not had any prior experience with building understanding, recording their thinking, or engaging in meaningful mathematical discourse. For classrooms where the opposite is true, the unit will go much faster as the culture of shared learning would have already been established. Any teacher who wishes to embark on this journey will have to move with the rhythm of her own classroom and follow the strategies and thinking that surface. This unit is meant to be a guide to help the teacher gain a bigger picture of how a child learns and what to expect, but every group of students varies in readiness. It is more important to follow your class than to stick too closely to the lessons in this book.

Common Core standards addressed
This book addresses the following two second grade common core standards:

> 2.NBT.B.5 – Fluently add and subtract within 100 using strategies based on place value, properties of operations, and/or the relationship between addition and subtraction.
> 2.NBT.B.9 – Explain why addition and subtraction strategies work, using place value and the properties of operations.

It also addresses all 8 of the standards for mathematical practices:

> Make sense of problems and persevere in solving them
> Reason abstractly and quantitatively
> Construct viable arguments and critique the reasoning of others
> Model with manipulatives
> Use appropriate tools strategically
> Attend to precision
> Look for and make use of structure
> Look for and express regularity in repeated reasoning

Build, Connect, Understand!

Historically, the methodology of teaching mathematics across the country has been that of traditional direct instruction. Using this method, teachers essentially demonstrate or model how to solve math problems and then provide opportunities for students to practice. Competence in mathematics is seen as being able to compute and use procedures quickly and accurately to solve math problems (Woodward, 2006). Researchers in mathematics education, however, would extend that definition by describing mathematical proficiency as: (a) conceptual understanding, (b) procedural fluency, (c) strategic competence, (d) adaptive reasoning, and (e) productive disposition (Kilpatrick, Swafford, & Findell, 2001). "Excessive use of transmission models of instruction undervalue the natural mathematical thinking children already use and communicate that in order to be mathematically proficient, a child must merely duplicate the thinking of the teacher" (Bahr, manuscript in progress). As a result, connecting students' prior knowledge to real-world mathematics through multiple representations and making explicit connections between them is considered to be most effective for teaching mathematics to all children (Solution Tree, 2010).

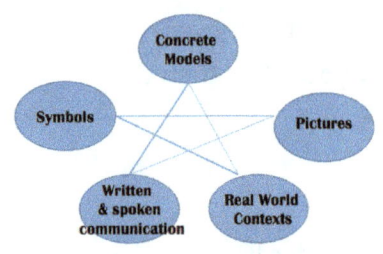

Lesh, Post, and Behr (1987) described five ways to represent mathematics: using symbols, pictures, concrete models, real-world contexts, and written or spoken language (Lesh, Post, and Behr, 1987). Schifter & Fosnot (1993) stated that, "teaching must be grounded in how students learn" (p. 193); therefore, being attuned to student thinking when helping to make connections between representations is important. These five types of representations vary in level of abstractness, and understanding how working within a level can either facilitate or hinder learning is important. Teachers tend to work with children on an abstract level; however, research has shown that when children are left alone to solve complex problems, they tend to act out the problem with some kind of a tool (Carpenter 1999). In other words, they naturally gravitate to the concrete as a way to have access to the problem.

The Freudenthal Institute for Teachers in the Netherlands has characterized on an iceberg model/metaphor three levels of abstractness: informal, pre-formal, and formal (Dekker, 2007). These levels can essentially be coupled with Carpenter's work, which concludes that when operating, students move from more informal or concrete strategies, such as directly modeling, to more formal or abstract ones such as using derived facts (Carpenter, 1999). This is further described by Hendrickson, Hilton,

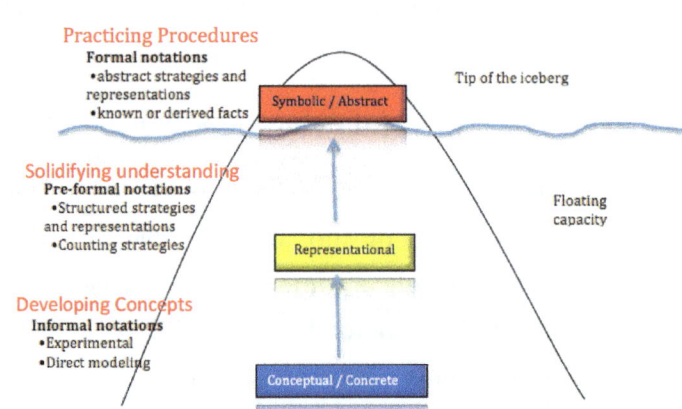

Figure 1. The iceberg model combining the ideas of Freudenthal Institute (2007) & Carpenter (1999) to show the strategies children use in each level of abstractness.

and Bahr (2009) as a *Learning Cycle* where students begin by developing concepts, then begin to solidify understanding as connections are made, before they are able to practice abstract procedures. These new procedures are then used as a basis to develop new concepts. Thus, the bottom of the iceberg comprises of concrete experiences targeting conceptual understanding, whereas the tip represents abstractness, utilizing symbolic notation. The center is focused on making connections between the concrete and abstract.

The movement toward the "tip of the iceberg" is supported by experiences that attend to the developmental readiness of students. Posing appropriate tasks, making available a variety of tools, encouraging students to communicate their understanding with their peers, and asking open-ended questions allow students to have access to solving the problem and to attend to the connections between representations.

This *Build, Connect, Understand* series is based on these principles. It takes a concept and looks at what is needed to build the understanding and how to make the appropriate connections to more abstract and numerical representations. Throughout the unit, we will use the terms "develop", "solidify", and "practice" to refer to where the lesson is approximately on the iceberg or within the learning cycle. Not every student will be solidifying their understanding at the same time, as some students will need more time to work with the concrete tools and develop concepts. However, the class as a whole will be moving along the continuum. Another thing to keep in mind in this addition and subtraction unit, there will be times where students will be solidifying or practicing addition, but in the broader picture of the entire unit, they may still be developing their understanding of more complex ideas.

The Lesson Map

An effective math lesson allows for opportunity for students to work as a whole group, small group, and by themselves. It embraces student discussion as a way to move the thinking in the classroom as well as emphasizes the importance of building understanding of concepts rather than just mimicking procedures shown by the teacher. Although each teacher may have her own way to finesse a lesson and guide her students to reach the goal of the particular lesson, the main components of a math lesson focused on understanding, or commonly referred to as an inquiry lesson, are launch, explore, discuss. Most teachers also engage students in a mathematical routine or warm-up unrelated to the main lesson for the day. The terminology of the lesson components may differ slightly from researcher to researcher, but each is critical and cannot be undermined or shortchanged without loosing the integrity of the lesson.

Launch

The launch is the part of the lesson where the teacher sets up the task to be explored. Typically, but not always, it is a smaller problem that will bring out some common understanding that students will need in solving a more complicated problem in the explore phase.

In the Launch, the stage is being set for learning and connections are being made to prior learning and previously developed strategies. These connections help build mathematical ideas. Examples of what might occur during the launch are:

- Summarizing the learning experiences of the students up to that point
- Posing an introductory activity to connect with prior knowledge or strategies
- Posing and clarifying expectations of the task to be explored and of issues related to management and discourse (Bahr & de Garcia, 2010 p 163)

In this unit, often there is an identical problem in the launch and explore phases, but the number sets are different. The initial problem uses smaller numbers, those that most 7 year-olds are able to conceptualize, while trying to wrap their heads around the complex language of the problem. Once the situation is made clearer, the numbers can increase. This prevents children from having two cognitive challenging things to deal with at once, language and number size. Typically, the launch is about 10 minutes, but may vary depending on the lesson.

Math Lesson Map
Planning a Lesson

Routines/Math Chat

A "warm-up" and/or a connection to a lesson (usually using number sense and operations, practice/review, and/or mental math)

Whole Group

Teachers plan effective lessons by determining what students will learn, identifying the big mathematical ideas that need to be taught, identifying potential misconceptions, and preparing necessary materials. Judge what is important depending on students' needs.

Launch

The teacher sets the stage for learning by ensuring the purpose and the rationale of the learning is understood by the students, connecting the purpose to prior learning, posing the problem(s) and clarifying the task for students, identifying the tools and materials available, and setting the expectations for the lesson (learning outcome, time, and structure).

The students understand the purpose, rationale and expectations of the learning

Whole Group

Explore

The teacher provides opportunities and support for students to develop conceptual understanding by providing meaningful explorations and tasks that promote active student understanding by providing meaningful explorations and tasks that promote active student engagement.

The teacher monitors the development of student understanding by observing student thinking and during questions to stimulate the thinking.

The student constructs meaning of the mathematical concept being taught and records math thinking in journals.

Individual, Pairs, or Small Group

Discuss

The teacher provides opportunities to make public the learning that was accomplished by the students by sharing the evidence of what was learned, providing opportunities for students to analyze, share, discuss, clarify, extend, connect, consolidate and record thinking strategies. A summary of the learning is articulated and connected to the purpose of the lesson.

The student is able to articulate the learning/understanding of the mathematical concept being taught through reflection in journal and communication with peers.

Whole Group

Homework
Teacher assigns meaningful practice to extend the learning

Explore

Once the task is introduced, students are to work with their partner, triad, or small group to solve the problem. Teachers may want to allow children some independent think time first so they have more to contribute to the discussion of the group, or decide that they all need to work together to get some ideas on paper. The groups need to be pre-determined and static for some time so that children have opportunities to get to know each other as mathematicians. Strategic pairing up of partners is critical in ensuring productivity.

The role of the teacher during the explore phase is to move from group to group to listen to what is being discussed, ask questions to help move the groups to another phase in their learning, determine if there are some common misconceptions, analyze the strategies that are surfacing so that she can determine which she may want to highlight later during the discussion piece, and facilitate strong partner work between the children.

The explore phase, depending on the age of the children and concept being explored, can last 10-30 minutes. Some problems are problematic enough to require more than one class session to work through.

Discuss

The discuss period is also referred to by Catherine Twomey Fosnot (2001) as the *math congress*. This is when the students reconvene as a whole group to discuss their findings and share their strategies like mathematicians. The teacher's role is to facilitate the discussion and select which strategies should be highlighted first that would lead to deeper mathematical understanding. This means that not all groups will be sharing. It is not a report, but a chance to get mathematical ideas out onto the table for all to consider. It is wise to choose a more

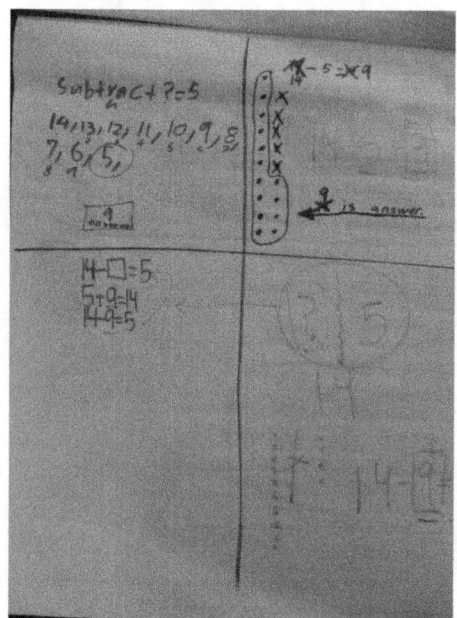

developmentally basic strategy, such as counting on, that most students would be able to resonate with before introducing one that is more novel and complex, such as subtracting. Teachers may choose to bypass an overly sophisticated or convoluted strategy that may work, but is beyond the understanding of most of the class. Allowing students during the explore phase to record their strategy on a chart paper divided into 4 sections, enables the class to make connections between them.

There should be at least 15-20 minutes dedicated to the discussion at the end of the lesson. This is the part that most teachers tend to skip due to time. However, this is the most critical part of the lesson, where the mathematical ideas come together, connections are made and algorithms are generated. Conjectures posed by students can be charted on a growing conjecture chart that is periodically challenged as new ideas surface. When made public, students take ownership of them and oftentimes the same students who made a conjecture will take the time to try to prove or disprove it.

Setting up an environment of discourse

Communication during the math lesson, commonly referred to as discourse, has been advocated by NCTM and other math professionals since before 1991 when the *Professional Standards for Teaching* was released. It is one of the five process standards: communication, reasoning and proof, problem solving, representation, and connections, although the other four also involve communication. The *Common Core* is also based on these process standards and communication is embedded throughout. Examples are its statements such as, "One hallmark of mathematical understanding is the ability to justify…a student who can explain the rule understands the mathematics and may have a better chance to succeed at a less familiar task…" (p. 4), and "Students also use the meaning of fractions, of multiplication and division, and the relationship between multiplication and division to understand and explain why the procedures for multiplying and dividing fractions make sense" (p. 33). These *Common Core* statements make it clear that conceptual understanding must be connected to the procedures, and that one way to deepen conceptual understanding is through the communication students have around concepts, strategies, and representations.

I have been incorporating the use of rich mathematical discourse for many years now and every year I am amazed as to how it has transformed children. Many students struggle with computation or may be more right-brained, visual, and concrete learners by nature. Using discussion that helps them to make connections is a wonderful way to help students make sense of what was once out of reach for them. Several of these students are social by nature, but math just does not make sense. Talking it out with their peers helps them to access and process the mathematics.

I have also seen gifted students, who tend to be independent and symbolic thinkers, become mathematically powerful after having to live in an environment where making connections between representations and explaining ones thinking was an expectation.

The culture in the classroom becomes more dynamic as the focus shifts from the student to the mathematics that is being explored. When the focus is placed on the mathematics and whether certain solutions and strategies make sense rather than what one person is saying, then students start to feel more comfortable and are more willing to throw out half an idea without fear of being criticized. They know that all their ideas are valued by the classroom community.

What is the role of the teacher in regards to facilitating rich mathematical discourse?

What teachers say and do can make or break a discussion. As outlined in the NCTM *Professional Standards for Teaching Mathematics* (1991 p. 35) teachers need to:

- Pose questions and tasks that elicit, engage, and challenge each student's thinking
- Listen carefully to students' ideas
- Ask students to clarify and justify their ideas orally and in writing
- Decide which of the ideas that students bring up during a discussion should be pursued in depth
- Decide when and how to attach mathematical notation and language to students' ideas
- Decide when to provide information, when to clarify an issue, when to model, when to lead, and when to let a student struggle with a difficulty
- Monitor students' participation in discussions and deciding when and how to encourage to participate

There are a lot of decisions that need to be made during a mathematical discussion, but when in doubt, ask a question. Some questions that can help get students to further explain their thinking are:

- How do you know?
- Why?
- Can you defend your thinking?
- What do you think about that?
- Does that make sense?
- Would someone like to add on to what s/he just said?
- Can you restate what s/he just said in your own words?
- How does that relate to....?
- How are ____'s strategy and ____'s strategy the same/different?
- Do you see a pattern?

Often times the best thing to do is nothing at all. Providing enough wait time can give students the right amount of processing time necessary to compose a thought. Initially, wait time might just be necessary because students are used to the teacher chiming in if no one answers a question. Hold your ground and wait for them to talk.

Setting up the right physical and emotional environment that supports discourse is critical. Probably the most effective seating arrangement, especially in the early stages, is for students to be facing each other by sitting in a U-configuration. Teachers can sit with their students as part of the circle to encourage peer-to-peer discussion. If teachers are still having difficult getting children to talk, they can remove themselves from the group and stand outside the circle. As a result, students are left looking only at each other, which encourages them to direct their comments to one another.

The emotional environment must be safe and be one where students want to learn and think deeply about the mathematics. By keeping a neutral tone and not praising, nor criticizing, the teacher is able to help focus the students' attention on the numbers rather than the person. For example, if a student solves 28 + 26 by adding 20 + 20 = 40; 8 + 6 = 15; 40 + 15 = 55, just record on the board what s/he says and ask the class what they think about that. Since this is something you would normally say, the class does not know if it correct or incorrect, so they are forced to match their thinking to the other student's strategy. Even the student who proposed the strategy would be looking over his / her own work to make sure s/he agrees with it. If a student catches the error and says that 8 + 6 = 14, tell the class that there is now a discrepancy and how they can prove which one is correct. Naming strategies after the students who posed them is a way to validate the child, yet keeping the focus on the math. Students feel good when they positively contribute to the conversation and the learning of others.

What is the role of the student when engaging in a mathematical discussion?

The student has a primary role during the discussion since they are the ones who have to be doing the talking. The NCTM *Professional Standards for Teaching Mathematics* (1991 p. 45) identifies the roles students play in the discourse as the following:

- Listen to, respond to, and question the teacher and one another
- Use a variety of tools to reason, make connections, solve problems, and communicate
- Initiate problems and questions
- Make conjectures and present solutions
- Explore examples and counterexamples to investigate a conjecture
- Try to convince themselves and one another of the validity of particular representations, solutions, conjectures, and answers
- Rely on mathematical evidence and arguments to determine validity.

Discourse is more complicated than just talking and listening. Children think that because they can hear one another they are listening, but listening involves thinking about what they are hearing. None of the preceding roles come naturally to children and need to be taught explicitly and continually.

What communication and social skills need to be taught for an effective mathematical discussion?

Students are not going to be great at discussing their thinking on day one. In fact, maybe not even on day 30, but it can improve a little every day if the teacher is explicit about teaching good communication and social skills. Each day before starting, remind students about specific behaviors you want to see and then provide feedback on those behaviors afterwards.

Initially, most all students need to learn eye contact, with their partner and the speaker during a whole group discussion. Small children need to be told to sit "knee to knee, eye to eye" or something similar, but even older students need to be reminded to "turn and talk" to their partner. Be explicit about what listening looks like. Let them know that even if some may be listening when they are not looking at the speaker, it appears that they are not listening and perceived as rude. Also, teach the speaker to wait until others are listening before sharing his/her ideas.

The skills that need to be taught will vary depending on your individual students. Think about how you would talk as a group of adults and measure how your students are performing compared to that. An example is hand-raising. Would you raise your hand in a gathering with your colleagues? Sometimes, but not usually. I personally teach my students during math discussions to not raise their hands, but for someone to start and to take turns like adults. They also need to learn how to let others talk and not dominate the conversation.

Oftentimes it is necessary for the teacher to select who will start and let the conversation go from there. What the students seem to need the most help with is to actually be engaged and listening to what the speaker is sharing. They should be matching their thinking to the ideas of the speaker to decide if what is said makes sense, matches his/her thinking, or perhaps is a much more efficient solution strategy. When other students are raising their hands as someone is sharing, that is an indication that they are thinking more about what they themselves want to say than what is currently being said, and I call them on it.

Using Problem Solving as a Vehicle for Understanding Computation

When assisting children to develop their understanding of math concepts, it was mentioned earlier that the more contextual and concrete the problem is, the more accessible it becomes. Give a kindergarten student a problem that involves division, but in the context of fair sharing, using small enough numbers, he will be able to solve the problem. However, give that child the same problem written numerically in a number sentence, and the problem all of a sudden becomes too complex to access.

In the book *Children's Mathematics: Cognitively Guided Instruction,* Carpenter et al (1999) identifies 13 different problem types related to addition and subtraction, as well as highlight the solution strategies children naturally use when solving these problems. Of his 13 problem types, the following ones are used in this Unit:

Problem type		Example
Join	result unknown	Ms. Cleveland sent 13 students to the library and Mr. Dugan sent 15 students to the library. How many students are at the Library?
	change unknown	I had 5 cookies in my cookie jar. I made some more. When I was finished, I had 12 cookies altogether. How many more cookies did I make?
Separate	result unknown	45 students were in the cafeteria. 23 students left to go onto the playground. How many students are still in the cafeteria?
	change unknown	Travis had 14 valentines. He gave some to his friends. Now he has 5 valentines. How many valentines did he give to his friends?
	start unknown	When the birthday party started, there were some party hats on the table. 8 children came and they each took a hat to wear at the party. Now there are 14. How many party hats were on the table to start?
	Compare: Difference unknown	How many more books did Mr. Dugan's class read than Ms. Patrick's class?

Carpenter also reveals the solution strategies that children use when attempting to solve such problems. When the problem type, number, or both, is too complex, children will resort to strategies that directly model the situation and will tend to build both number sets. Children who are more comfortable with the number size might use strategies such as adding on, using a number line, or other counting strategies. If the student is comfortable with the problem type and the numbers, he will use more sophisticated strategies such as using number facts or a numerical procedure. Changing the problem type and numbers alter the complexity of the problem. By observing the students' approach the problem, the teacher will be able to determine where they are developmentally and can help him reach his next level.

Direct modeling	Counting	Number Facts
Result unknown	Change unknown	Start unknown

The diagram above shows the progression of strategies and problem types. When a child is struggling, and is directly modeling, then the teacher needs to realize that as some connections are made, the student will start to use counting strategies first before a formalized operation. Take this problem into consideration:

> I had 5 cookies in my cookie jar. I made some more. When I was finished, I had 12 cookies altogether. How many more cookies did I make?

A child trying to wrap his brain around this problem type will start with 5 cubes, then add more until he makes 12. When this same problem type is familiar, he will realize that he just needs to take 5 from 12 and thus use subtraction.

If a harder problem type is posed, then the solution strategy is sure to go down. A child solving a result unknown problem might use number facts, but when posed with a change or start unknown problem might start directly modeling it. Therefore, it will be most supportive to lower the number complexity at this time.

The goal is for children to understand the action of the numbers and to figure out a way to record it that makes sense. Once they are quite comfortable with this process, they can begin to refine their recording strategies to more efficient ones, such as the traditional algorithm, with full comprehension of what the markings mean.

Basically, I tell my student that they are not allowed to use a strategy that they do not understand. If they want to use it, they need to figure out why it works.

Math makes sense!

Two-digit Addition and Subtraction with Regrouping Unit Overview

Practice
Solidify
Develop

1 - Groups and Leftovers	2 - Representing a number	3 – Join Result Unknown	4 - Join Result Unknown	5 - Separate Result Unknown
D	D	D	D	D
6 – Separate Result Unknown	7 – Join Result Unknown regrouping	8 – Join Result Unknown – regrouping	9 – Join Result Unknown regrouping	10 – Join Change Unknown
D	D	D	D	D
11 – Separate Change Unknown	12 – Separate Change Unknown	13 – Separate Start Unknown	14 – Trying on Thinking	15 – Trying on Thinking
D	D	D	S	S
16 – Trying on Thinking	17 – Using Graphs	18 – Using Graphs Comparing	19 – Using Graphs Comparing	20 – Using Graphs Comparing
S	S	S	S	S
21 – Using Graphs Comparing	22 – Trying on Subtraction	23 – Trying on Subtraction	24 – Answering questions about data	25 – Answering Questions about Data
S	S	S	S	S
26 – Solidifying and Practicing Subtraction	27 – Practicing Subtraction	28 – Practicing Subtraction	29 – Practicing Subtraction	30 - Addition and Subtraction Assessment
S/P	P	P	P	P

15

Two-digit Addition and Subtraction with Regrouping
Daily snapshot

1 - Groups and Leftovers	2 - Representing a number	3 - Join Result Unknown	4 - Join Result Unknown	5 - Separate Result Unknown
		Ms. Cleveland sent 13 students to the library and Mr. Dugan sent 15 students to the library. How many students are at the Library? (24, 12)	Ms. Cleveland send 31 students to the computer lab and Mr. Dugan sent 18 students to the computer lab. How many students are at the computer lab?	45 students were in the cafeteria. 23 students left to go onto the playground. How many students are still in the cafeteria?
6 – Separate Result Unknown	7 – Join Result Unknown regrouping	8 – Join Result Unknown – regrouping	9 – Join Result Unknown regrouping	10 – Join Change Unknown
37 students were in the cafeteria. 24 students left to go onto the playground. How many students are still in the cafeteria?	Ms. Cleveland sent 28 students to the library and Mr. Dugan sent 16 students to the library. How many students are at the library.	The bucket of cubes fell on the floor. In one handful I collected 39 cubes. In another handful I collected 43 cubes. How many cubes did I pick up from the floor?	I went outside to collect cherry blossoms. In one bag I put 28 blossoms, and in another bag I put 36. How many blossoms did I collect? (37 & 19)	I had 5 cookies in my cookie jar. I made some more. When I was finished, I had 12 cookies altogether. How many more cookies did I make? (15, 32)
11 – Separate Change Unknown	12 – Separate Change Unknown	13 – Separate Start Unknown	14 – Trying on Thinking	15 – Trying on Thinking
Travis had 14 valentines. He gave some to his friends. Now he has 5 valentines. How many valentines did he give to his friends? (41, 18)	Jonathan made 34 snowballs. Some of them melted. He now has 19. How many of his snowballs melted?	When the birthday party started, there were some party hats on the table. 8 children came and they each took a hat to wear at the party. Now there are 14. How many party hats were on the table to start? (27, 29)	38 + 17 using a strategy that surfaced in day 13's lesson.	Solving problems using one or more of the strategies that have surfaced. 36 + 25 29 + 46 16 + 37 52 + 48

16 – Trying on Thinking	17 – Using Graphs	18 – Using Graphs Comparing	19 – Using Graphs Comparing	20 – Using Graphs Comparing
Continuing to solve problems using one or more of the strategies that have surfaced. 55 + 26 72 + 19 38 + 43 58 + 43	Writing and solving questions generated by data on a graph.	Continue to use the data on the graph to answer questions that compare data. How many more books did Mr. Dugan's class read than Ms. Patrick's class?	Trying out a subtraction strategy.	Continue to use the data on the graph to answer questions that compare data. How many more books did Mr. Dugan's class read than Ms. Patrick's class?
21 – Using Graphs Comparing	22 – Trying on Subtraction	23 – Trying on Subtraction	24 - Answering questions about data	25 – Answering Questions about Data
Using expanded notation, practice using renaming when trading.	Trying on renaming as a strategy when recording subtraction of blocks. 34 - 16	Continue to try on the renaming strategy. 43 – 17 34 – 18 65 – 36 51 - 36	Using additional data sets on previous graph so it is now a double bar graph, students consider questions that can be asked and answer them.	*How many more books did Ms. Patrick's class read in Q1 than in Q2? Give a possible reason that her score may have increased so much. Name whatever strategy you chose to use
26 – Solidifying and Practicing Subtraction	27 – Practicing Subtraction	28 – Practicing Subtraction	29 – Practicing Subtraction	30 - Addition and Subtraction Assessment
Provide an erroneous strategy that students have to analyze whether or not it works. 50 – 27 = 37	Provide another erroneous strategy to analyze, then provide practice problems.	Using work from day 27, create small groups to support where students are developmentally with subtraction. Help them to either clean up their recording to a more efficient one, or to make stronger connections by continuing to build the problems.	Continue to work in small groups as needed to help students move into their next phase. (Take additional time as needed).	

Groups and Leftovers

Day: 1

Objective: In the context of creating groups with counters, students will see the connection between bundling in groups and left overs to the written two-digit numeral, thus understanding the significance of grouping into tens and ones.

Materials: Counters, individual recording charts, and an enlarged version of the recording chart for 6 different numbers on chart paper or interactive white board.

Launch: Provide counters to partnerships. Ask all partnerships to get 32 counters and put them into groups of three. On their recording sheet, they are to record how many groups of threes they were able to make and how many left overs. They are to do this then for 3, 5, 7, 8, and 10. Bring the group together and record on the chart paper the different numbers of groups and left overs made with each number. Discuss the process to make sure students understood how to do the activity and iron out any difficulties with the recording.

Explore: Pass out a sticky note to each partnership with a 2 digit number written on it. Note if certain partnerships need a larger or smaller number and hand out accordingly. Students will use the same method as with the previous number (32). When they finish, they can record their work on one of the class charts. Students who are quick to finish can be assigned an additional number. It is only necessary to collect data from about 5 additional numbers.

As you walk around, notice which students are having trouble making groups or recording the groups and left-overs on the recording sheet.

Discuss: Have students stand back and see what patterns they notice when looking at all the charts on the wall. They will find some interesting patterns, but ultimately you want to make sure they see that the group with the tens and left overs is the same as the written numeral. If they do not see that on their own, after discussion of other patterns is exhausted, circle the bottom row on each chart (10's) and ask what they notice with the numbers in that row.

Exit Slip: Ask students to write the number that has 3 tens and 9 ones.

Groups and Leftovers

	Groups	Left-overs
Number in each group		
3		
5		
7		
8		
10		

	Groups	Left-overs
Number in each group		
3		
5		
7		
8		
10		

Class Notes:

Representing a Number

Day: 2	**Objective:** Students will represent numbers using base ten blocks and/or unifix™ cubes as tens and ones and be able to draw a pictorial representation on paper.
Materials:	Base ten blocks, unifix™ cubes, journal
Launch:	After a brief discussion to recap what students discovered during the grouping lesson from the previous day, ask students to use materials provided (base ten blocks and unifix™ cubes) to represent the number 24. Students will build 24 in a variety of ways, such as with 24 ones, 24 tens, 2 tens and 4 ones you get multiple representations, such as 24 ones, 24 tens, and 2 tens and 4 ones, take some time to discuss how they are the same and different and which is the most efficient way to represent the quantity. Note: it is more than likely that there are students actually "drawing" the number with the cubes. Do not say anything. The class needs to debate which is a true representation of the number you posed.
Explore/ Discuss: 1	Pose another number and ask students to both build it and draw it in their journal. The point of this lesson is for students to begin recording 2 digit numbers pictorially. Notice if students are still struggling with representing the quantity. Also, notice the variety of ways students are drawing. Have a couple of students share out their recording and prove that it matches what they built. Choose one student who may have drawn a large picture with a lot of detail and one who drew a very efficient picture of sticks and dots. Have the class compare the drawings and decide which is most efficient. Students will comment on things like it taking less space and not using so much paper, or taking too much time to draw all the little lines of the blocks.
Explore/ Discuss: 2	Repeat using another number or two as necessary for all students to be on the same page with the representations. If all the students are still putting in too much detail, as they draw another number, tell them that you are having a contest to see which is the most efficient and include yourself. Draw the tens and ones with small sticks and dots and when all students show their work for all to see, they will clearly see that yours is the most efficient. ❘❘❘ •.
Exit Slip	Have students draw the number 45 using a pictorial representation.

Excerpt from the classroom:

I always make sure that I add this lesson before exploring addition and subtraction with regrouping just to make sure that we are all on the same page with how to use base ten blocks to represent a number. I do this because several years ago I was in a classroom and by chance I asked second grade students to build the number 24. What I got was a lot of "drawings" of 24 with the base ten blocks, but they did not represent the quantity of 24. Now days, I see that less and less, but in the upper grade classrooms I will still see that if I ask students to show me what 3 x 4 looks like. They will put 3 cubes, cross two ten sticks, add 4 more cubes and then put two parallel ten sticks to represent an equal sign and then show 12 cubes on the right.

Class Notes:

Join Result Unknown
no regrouping & no recording

Day: 3

Objective: In the context of solving a join result unknown problem, students will solve for an addition problem using a variety of tools. The teacher will notice the solution strategies and tool choice of the different children.

Note: The focus for today is problem solving using the tools and monitoring thinking. Do not yet focus on recording numerically.

Materials: Unifix™ cubes, base ten blocks, hundred charts, number lines (retractable measuring tapes), cubes and cups (for tens)

Launch: Pose the following problem on the board (chart paper or interactive board) –substitute names of teachers in your school.

   Ms. Cleveland sent 13 students to the library and Mr. Dugan sent 15 students to the library. How many students were sent to the library?

Allow students to work in partnerships and solve the problem using one of the tools available. Set clear expectations of what you expect to see while students are working in partnerships.

When the students are done, discuss how they approached the problem. Quickly share 2 strategies, starting with those who may have used a direct modeling (building all) and build up to children who may have used the 100 chart or tens and ones as a tool. If students are not used to explaining their thinking, they may not have paid attention as to how they solved the problem. If this is the case, make that the focus of the next problem.

Explore: Pose the same problem, but change the numbers:

 Ms. Cleveland sent 24 students to the library and Mr. Dugan sent 12 students. How many students were sent to the library now.

Observe children to see how they are approaching this problem. Some will be counting and building all, some will be counting on, and some will trade to make an extra ten.

Discuss: Ask students to share how they solved their problem, this time with more attention as to how they got the answer with their tools and not just representing it with the blocks. It is most helpful when sharing out if children are seated in a circle on the floor so all able to see each other's model. If students have not developed a culture of communication in the classroom, the true focus of the discussion is how to listen when a classmate is sharing his/her idea. Try to find strategies that are similar and help children make connections between them. For example, if a student grouped ten cubes in a cup to make a ten and another student used base ten blocks, have them talk about why those are similar. If a student just counted on with their tens and ones and another student used a number line, they might discuss how the two strategies they are connected. You may want to model how to draw the thinking of the students, but refrain from representing that with numbers at this point.

Exit Slip None

Excerpt from the classroom:

In the room where I taught this lesson, the struggle the students had was that they didn't attend to how they solved the problem. Some students were able to solve the problem mentally and then represented it with tens and ones. Two students combined unifix™ cubes and base ten blocks and we discussed that we needed to be working with the same unit because ten of one object is different than ten of another. Others used a hundred chart or measuring tape by counting on. This is why we are not focusing in recording. Students are taxed enough by trying to pay attention to how they solve a problem and try to articulate it with the rest of their classmates. During the share out, one student showed how on the hundred chart she used small cubes to count and placed 2 cubes and 1 cubes on the hundred chart and got "3", removed them and placed 4 and 2 to get "6". She then stated that she had 36. When students were asked to talk to their partners about that, their reactions were mixed, some thinking that it was "smart" and others saying how that was just 9 and not 36.

Class Notes:

Join Result Unknown
No regrouping & initial recording with pictures

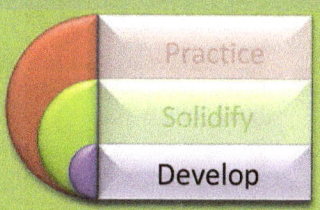

Day: 4	**Objective:** In the context of solving a join result unknown problem, students will solve for an addition problem using a variety of tools and to continue to monitor their thinking as they solve the problem so they can share their strategy with a partner.
Materials:	Unifix™ cubes, base ten blocks, hundred charts, number lines, cubes and cups (for tens)
Launch/ Explore:	Pose the following problem on the board (chart paper or interactive white board) –substitute names for teachers in your school. Ms. Cleveland sent 31 students to the library and Mr. Dugan sent 18 students to the library. How many students are at the library? Allow students to work in partnerships and solve the problem using one of the tools available. Set clear expectations of what you expect to see while students are working in partnerships, such as both partners working and solving the problem with the tool and not just representing the answer. As students finish, hand them a recording sheet so that they can try to write their strategy on paper. This part is quite challenging if students are not used to documenting their thinking. Carefully monitor the students and ask them questions to help them more clearly show the process they used to get the solution.
Discuss:	The task today is actually trying to get the thinking on paper, therefore the discussion should be how they accomplished this. Pick a tricky area, such as using the number line and compare students' recordings, starting with a basic one, and one that clearly shows how the child counted. Encourage all students to help come up with an even more efficient way to use the tool and to record their thinking on paper.
Exit Slip	None

Excerpt from the classroom:

The students were better at attending how they solved the problem, but struggled with how to record it on paper. When using either Unifix™ cubes or base ten blocks, most students had simply recorded the answer using 4 lines and 9 dots, or something similar. One student had actually considered how he put them together. He drew arrows to show the movement, and circled the ones to show that he moved all of them. When conferring with students who simply showed the answer, I let them know that I was unable tell by the picture how they put their cubes together, and asked them to think about how they could show that.

Another difficult tool to record was the number line. Many students had written all the numbers from 1-49. However, there was no representation on how they actually co I asked them if they had counted 1…2…3…4…etc., to which they replied no, they had started at 31 and counted on 18. Their task was to figure out how to show that pictorially. One little girl had already come up with that before I even approached her. Her representation looked like:

For her initial attempt, it was very clear. At the final discussion, I had this student share her work on the document camera and the students had to make sense of her recording. We also talked about other ways they could have counted up besides by ones, since this is how all the students counted.

Class Notes:

Separate Result Unknown
No regrouping – recording w/ pictures

Day: 5	**Objective:** In the context of solving a join result unknown problem, students will solve for a subtraction problem using a variety of tools and will continue to monitor their thinking as they solve the problem so they can share their strategy with their peers.
Materials:	Unifix™ cubes, base ten blocks, hundred charts, number lines, cubes and cups (for tens)
Launch:	Pose the following problem on the board (chart paper or interactive white board). *45 students were in the cafeteria. 23 students left to go onto the playground. How many students are still in the cafeteria?* Allow students to work in partnerships and solve the problem using one of the tools available and then record their thinking on the recording form. Set clear expectations of what you expect to see while students are working in partnerships, such as both partners working and solving the problem with the tool and not just representing the answer. Carefully monitor the students and ask them questions to help them more clearly show the process they used to get the solution. Remind them of other students' thinking that was discussed the previous day to see if they might want to try on a different recording strategy. This is the first time students will be recording subtraction, so help them think through how they can show in a picture what they did with the blocks, or other tools.
Discuss:	Again, the discussion is not about the answer but how they are able to show their thinking on paper. This is important so when they begin using numbers that require regrouping, they have an idea on how to show action on paper. Also, discuss efficient recording or solving strategies with the 100 chart or number line. Are there students who are not counting on or writing every number on the number line?
Exit Slip	None

Excerpt from the classroom:

Recording today was much better. The challenge was how to show subtraction on paper. Many students drew a representation like:

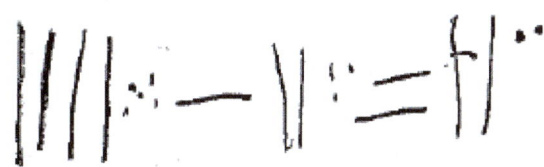

When they do that, ask students to show you what they did with the blocks to find the answer and they will sure to show you that they removed two tens and three ones. Tell them that their challenge is to figure out what they could do on paper to show that. They will be able to come up with a crossing out method fairly easily. Explain that the addition and subtraction symbols go with numerals and they can show action with pictures in a different way.

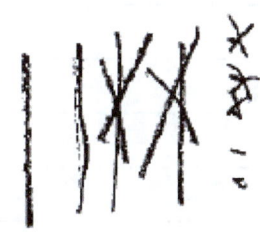

We had a variety of strategies on the number line as well. Many students again wrote all of the numbers until 45 on their line, but some students had other ways to record their thinking. This student started with a number higher than 0 to make sure that she would have enough room, and was the only one who thought of counting back by tens.

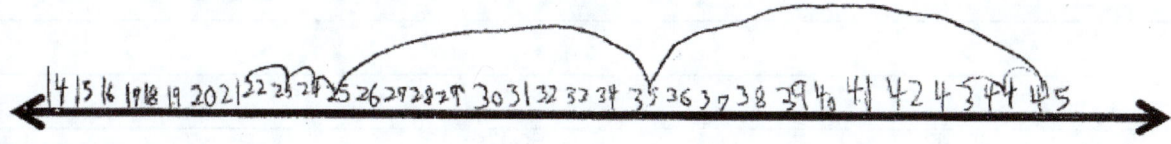

This student made increments of 5 and then counted back by 5's.

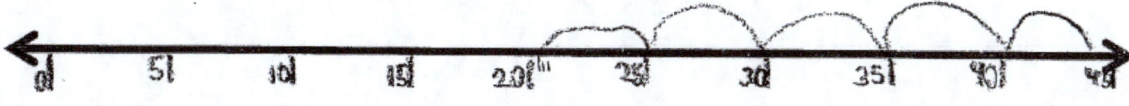

Class Notes:

Separate Result Unknown
No regrouping – recording w/ pictures

Day: 6	**Objective:** In the context of solving a join result unknown problem, students will solve a subtraction problem using a variety of tools. They will continue to monitor their thinking as they solve the problem so they can share their strategy with peers.
Materials:	Unifix™ cubes, base ten blocks, hundred charts, number lines, cubes and cups (for tens)
Launch:	Pose the following problem on the board (chart paper or interactive board). *37 students were in the cafeteria. 24 students left to go on the playground. How many students are still in the cafeteria?* Allow students to work in partnerships and solve the problem using one of the tools available and then record their thinking on the recording form. Carefully monitor the students and ask them questions to help them more clearly show the process they used to get the solution. Remind them of other students' thinking that was discussed the previous day to see if they might want to try on a different recording strategy.
Discuss:	Again, the discussion is not about the answer but how they are able to show their thinking on paper pictorially. This is important so when they begin using numbers that require regrouping, they have an idea on how to show action on paper. Also, discuss efficient recording or solving strategies with the 100 chart or number line. Are there students who are not counting on or writing every number on the number line? At this time, we are not yet connecting their strategies with any formal notation or algorithm.
Exit Slip	None – keep recording sheets

Excerpt from the classroom:

The students have been taking on the strategies of their friends. Today many counted back by tens and ones on the hundred chart as well as the number line. One student even tried counting back by tens and ones on a number line. She marked the top with tens and the bottom with ones. She started on 30 and counted back 20 and landed on 10, then counted back 4 from 7.

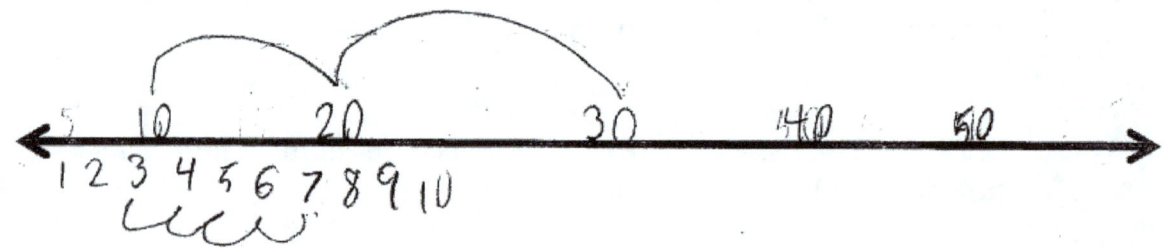

Another student started at 37 and counted down until she reached 24, but did not exactly know where her answer was, and we had to work on keeping track of her jumps.

Class Notes:

Join Result Unknown
Regrouping Recording with pictures and initial recording with numbers

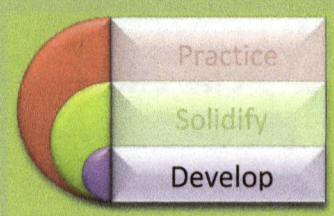

Day: 7	**Objective:** In the context of solving a join result unknown problem, students will solve for an addition problem that requires regrouping using a variety of tools. They are continuing to consider how to represent their thinking in pictures and how to record that thinking most efficiently.
Materials:	Unifix™ cubes, base ten blocks, hundred charts, number lines, cubes and cups (for tens)
Launch/ Explore:	Pose the following problem on the board (chart paper or interactive white board) –substitute names for teachers in your school. Ms. Cleveland sent 28 students to the library and Mr. Dugan sent 16 students to the library. How many students are at the library? Allow students to work in partnerships and solve the problem using one of the tools available. Have them record their thinking on the recording sheet. Today will be tricky since trading for a ten is involved and students are not yet used to how to show this kind of action pictorially. Carefully monitor the students and ask them questions to help them more clearly show the process they used to get the solution.
Discuss:	Again, there will be no question as to what the solution is, but the emphasis in the discussion will be how different students chose to represent their thinking. If a student put the tens together and counted on the ones, have students come up with how that should be represented. Most students will have taken ten of the ones and traded it for a ten stick, if using base ten blocks, so ask how that can be most efficiently shown in the picture. Simply drawing both numbers and then a final answer doesn't show the reader how the student came up with that number.
Exit Slip	None

Excerpt from the classroom:

Today's challenge was by far trying to get the action of the trade on paper. By far, most students recorded their problem this way:

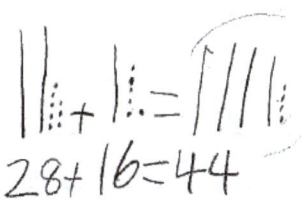

The role of the teacher at that moment was to ask the student how s/he knew that 44 was the answer, or how they arrived with that picture since only 2 tens and 1 ten can be seen in the problem. As the child acts out how they used the blocks to solve the problem, then encourage him to figure out a way to show the strategy that was shared on paper. Then they will able to come up with some kind of grouping strategy such as:

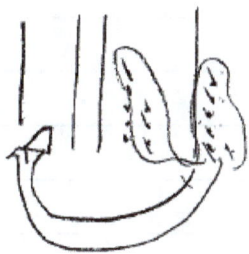

The children's number lines are also improving. Today a student came up with quite an efficient strategy for recording. This was an important point in the discussion since some children are still struggling with writing every number.

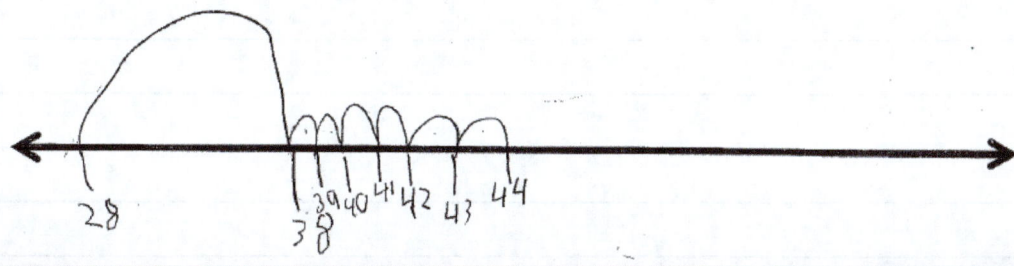

Class Notes:

Join Result Unknown
Regrouping Recording with pictures and initial recording with numbers 2

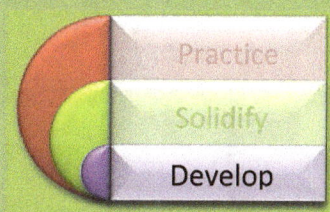

Day: 8	**Objective:** In the context of solving a join result unknown problem, students will solve an addition problem that requires regrouping using a variety of tools. They are continuing to consider how to represent numerically what they are doing with the blocks.
Materials:	Unifix™ cubes, base ten blocks, hundred charts, number lines.
Launch/ Explore:	Pose the following problem on the board (chart paper or interactive board)

The bucket of cubes fell on the floor. In one handful I collected 39 cubes. In another handful I collected 43 cubes. How many cubes did I pick up from the floor?

Allow students to work in partnerships and solve the problem using one of the tools available. Have them record their thinking on the recording sheet. Today encourage the students to pay attention to what they are doing with the blocks to see if they can make a numeric representation of that action. |
| **Discuss:** | Again, there will be no question as to what the solution is, but the emphasis in the discussion will be on how students are recording numerically. If students are still not taking this on, stop the lesson a little early, after they have solved it and drawn a picture, and then have a student share what they did with the blocks to add the numbers above together. Record first pictorially what the student said (since many will have solved it in a very similar fashion), then think out loud how you might record that thinking in numbers in a couple of different ways. |
| **Exit Slip** | None |

Excerpt from the classroom:

The students today were able to consider how to draw the actions of combing their cubes together, but not yet ready to think about the numbers. We had a "mid-workshop share" where I had to pull the class together so we could discuss how we might do this. One student shared a typical strategy of combining the blocks, by taking one cube to make a ten with the 9, then trade it for a ten. I modeled three different ways to record this thinking:

$$39 + 43$$
$$40 + 42 = 82$$

$$39 + 43$$
$$70 + 10 + 2 = 82$$

$$39 + 43$$
$$30 + 40 = 70$$
$$9 + 1 = 10$$
$$70 + 10 + 2 = 82$$

The emphasis is on having the students attend to what they are doing and making numeric notations of this action. They are not yet ready to consider how this would look using a standard algorithm. It is more important for them to get their thinking down on paper in a way that makes sense to them at this point.

Class Notes:

Join Result Unknown
Regrouping - Recording with pictures and numbers

Day: 9	**Objective:** In the context of solving a join result unknown problem, students will solve for an addition problem that requires regrouping using a variety of tools. They are continuing to consider how to represent numerically what they are doing with the blocks.
Materials:	Unifix™ cubes, base ten blocks.
Launch/ Explore1:	Pose the following problem on the board (chart paper or interactive board). *The bucket of cubes fell on the floor. In one handful I collected 28 cubes. In another handful I collected 36 cubes. How many cubes did I pick up from the floor?* Allow students to work in partnerships and solve the problem using one of the tools available. Have them record their thinking in their math journals. Today they will focus on the base ten blocks (and using unifix™ cubes if needed). Today encourage the students to pay attention to what they are doing with the blocks to see if they can make a numeric representation of that action.
Discuss1:	After student have spent some time on this problem, they may have descent pictures of what they are doing to solve it, but not yet using numbers. Gather the students and have someone share out his/her thinking and record it pictorially and then model how you would record using numbers. Have a couple more students share out so that the two or three most common strategies are shared and recorded. Have students choose which method matched their way and to copy the recording strategy. This will help children try on one recording strategy. It is probably that they will use the same strategy for the next problem.
Explore2:	Pose the following problem on the board (chart paper or interactive board) and have students solve with their partner an enforce trying to use numbers to record their actions and picture. *The bucket of cubes fell on the floor. In one handful I collected 37 cubes. In another handful I collected 19 cubes. How many cubes did I pick up from the floor?*

Discuss2:	After an adequate amount of time solving this problem, refer to one of the problems shared out earlier and ask who tried on that particular strategy. Have the student show his/her work, or explain it in words and the teacher can record. Do this for the most important strategies that have surfaced.
Exit Slip	Look at recording in journals to see who were able to write a numeric representation versus those who did not. Check to see if the representation matched their picture or if it was a distinct strategy.

Excerpt from the classroom:

This was a breakthrough lesson where students were finally able to start to show their thinking in numbers as well as pictures. There were three students who were not yet writing their thinking in numbers, so those are the ones we will be monitoring tomorrow.

Class Notes:

Join Change Unknown
Recording with pictures and numbers

Day: 10	**Objective:** In the context of solving a join change unknown problem, students will use multiple representations to make sense of the problem. Students will also start to investigate the use of inverse operations in solving a problem.
Materials:	Unifix™ cubes, base ten blocks, number lines
Launch/ Explore1:	Pose the following problem on the board (chart paper or interactive white board). *I had 5 cookies in my cookie jar. I made some more. When I was finished, I had 12 cookies altogether. How many more cookies did I make?* Allow students to work in partnerships and solve the problem using one of the tools available. Have them record their thinking in their math journals. This is the children's first exposure to a join change unknown problem, the purpose of this first question is to make some sense of the problem and not complicate it yet with difficult numbers. As you walk around the room, help question the students to guide them to make as clear of a representation as possible.
Discuss1:	After an adequate amount of time solving this problem, have students share out different ways that they solved it. Start with strategies such as adding up or counting up on a number line, since it more directly mirrors the problem. Move towards other representations that might include subtraction.
Explore2:	Pose the following problem on the board (chart paper or interactive white board). *I had 15 cookies in my cookie jar. I made some more. When I was finished I had 32 cookies altogether. How many more cookies did I make?* Again, allow students to work with their partner to solve this more challenging problem. Encourage them to draw a representation that is clear and to represent their strategy numerically.
Discuss2:	After an adequate amount of time solving this problem, have students share out their different strategies. Start with ones that mirrored how students solved the first problem so that they are able to connect what was done with the smaller numbers to larger ones.
Exit Slip	Students hand in journals to make sure that they have a pictorial and numerical representation of their thinking of the second problem.

Excerpt from the classroom:

When posing a problem, especially if it is different than the ones with which students are previously familiar, it is important that the teacher takes time to think of as many ways to solve it, or as many ways they think children would solve it. In the case of the first problem, here are some possible solutions:

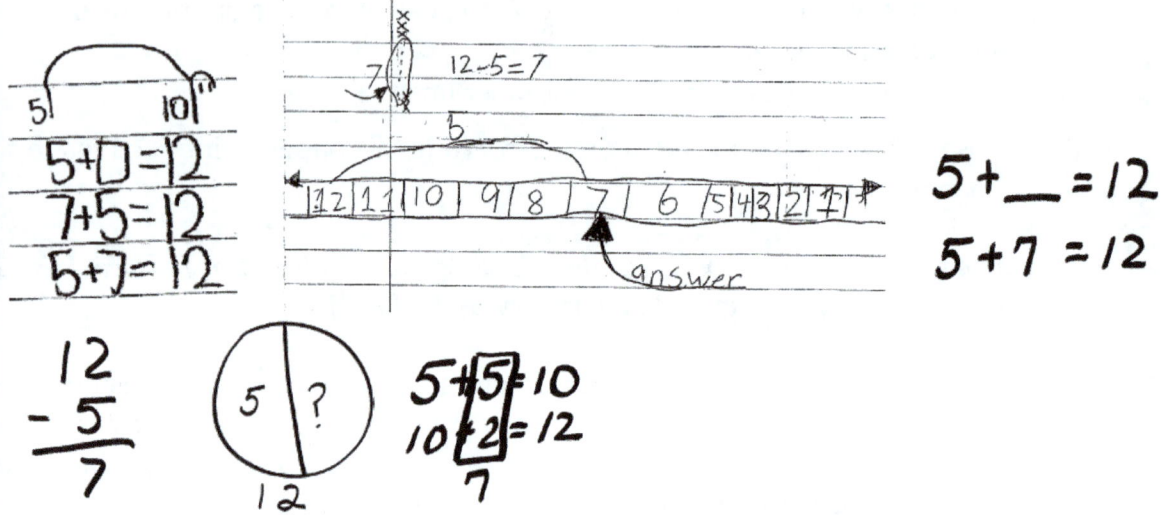

Since the second addend is unknown, those who are proficient in math would go straight to subtracting the amount of cookies that were in the jar in the beginning from the amount that were there at the end. However, children who are just building their understanding around addition and subtraction, will take the problem more literal and solve it by either adding on or using trial and error.

It is important that students have some time to wrap their brains around this new problem type with very simple numbers. Using smaller numbers enables the child to conceptualize what is happening in the story better than if there were complicated numbers. This way they can check their answers using reasonableness.

As the numbers increase in size, some children resort to taking the two numbers and adding them, not thinking about the context. Help those students act out the situation with their cubes and a bag. The first problem left the class with a bank of strategies that can be applied to larger numbers. These larger numbers, however, allow for even more strategies, such as:

Class Notes:

Separate Change Unknown

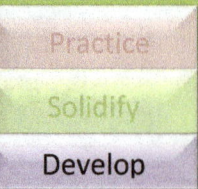

Day: 11	**Objective:** In the context of solving a separate change unknown problem, students will use multiple representations to make sense of it. Students will also continue to investigate the use of inverse operations in solving a problem.
Materials:	Unifix™ cubes, base ten blocks, number lines
Launch/ Explore1:	Pose the following problem on the board (chart paper or interactive board). *Travis had 14 valentines. He gave some to his friends. Now he has 5 valentines. How many valentines did he give to his friends?* Allow students to work in partnerships and solve the problem using one of the tools available. Have them record their thinking in their math journals. This is the class's first exposure to a separate change unknown problem, the purpose of this first question is to make some sense of the problem and not complicate it yet with difficult numbers. As you walk around the room, question the students to guide them to make as clear of a representation as possible. Encourage them to directly model the problem or act it out.
Discuss1:	After an adequate amount of time solving this problem, have students share out different ways to solve it. Start with strategies such as adding up or counting up on a number line, since it more directly mirrors the problem. Move towards other representations that might include subtraction. Ask students to compare this problem with the one from day 10.
Explore2:	Pose the following problem on the board (chart paper or interactive white board). *Alexis had 41 valentines. She gave some to her friends. Now she has 18. How many valentines did she give to her friends?* Again, allow students to work with their partner to solve this more challenging problem. Encourage them to directly model it and draw a representation that is clear and to represent their way numerically.
Discuss2:	After having an adequate amount of time to solve this problem, have students share out their strategies. Start with strategies that mirrored what was done in the first problem so that they can connect what was done with the smaller numbers to larger ones.
Exit Slip	Students hand in journals to make sure that they had both a pictorial and numerical representation of their thinking of the second problem.

Excerpt from the classroom:

The children are getting better at recording their thinking and they are trying to figure out how to record in numbers what they build. When presenting the first problem, some students built 14 and took some away until they had 5, and others built 14 and took away 5 to reveal how many valentines had been given out.

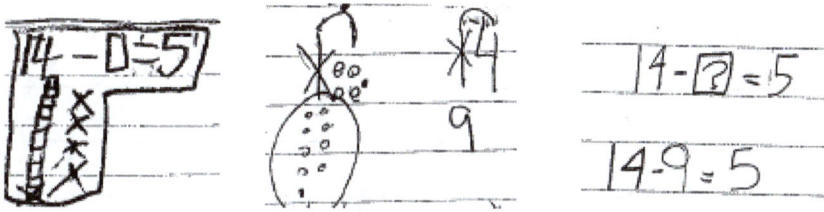

Some students used number lines to solve the problem and used one of these ways as well, either starting at 14 and counting back to get to 5, or starting at 14 and counting back 5.

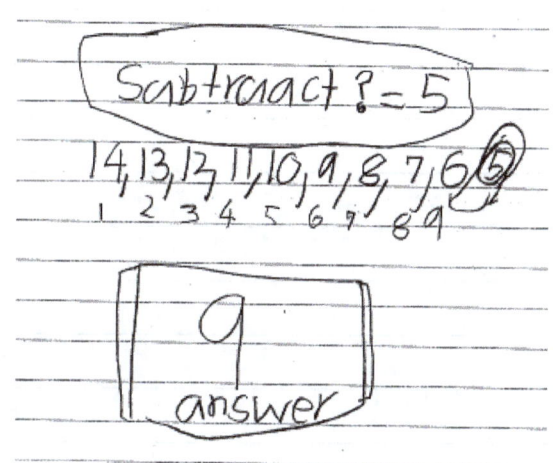

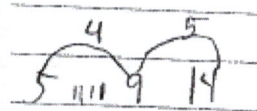

It was interesting, however, that there were a couple of students who somehow took away the ten and added the four and five together to get 9. I think they knew the answer and found a convenient way to make it, although it didn't have anything to do with the story.

The bottom line with the different strategies is to look at the different number sentences that were created to solve the problem: 14 – 5 = 9 or 14 – 9 = 5 to see how they are related.

When discussing student strategies from the second problem, a lot of time was focused on which number sentence would match a specific strategy, 41 – 18 = ___, or 41 - ___ = 18.

Class Notes:

Separate Change Unknown
Another look

Day: 11	**Objective:** In the context of solving a separate change unknown problem, students will use multiple representations to make sense of the problem. Students will also continue to investigate the use of inverse operations in solving a problem.
Materials:	Unifix™ cubes, base ten blocks, number lines
Launch/ Explore:	Pose the following problem on the board (chart paper or interactive white board). *Jonathan made 34 snowballs. Some of them melted. He now has 19. How many of his snowballs melted.* Again, allow students to work with their partner to solve. Encourage them to directly model it and draw a representation that is clear and to represent their way numerically.
Discuss:	Focus on the strategy students are using with the base ten blocks. Most all students will have traded a ten for ones, taken away 9 and then taken away ten. As a class discuss how they can record this numerically.
Exit Slip	Students hand in journals to make sure that they have a pictorial and numerical representation of their thinking of the second problem.

Excerpt from the classroom:

Students are still trying to wrap their heads around how to record with numbers the way they are subtracting with the base-ten blocks. Therefore, it is important to not yet use shortcuts. If they are building 34 with 3 tens and 4 ones, then the number should be recorded as 30 and 4. When they trade, they now have 20 and 14. Then they subtract ones from 14 and a ten from 20.

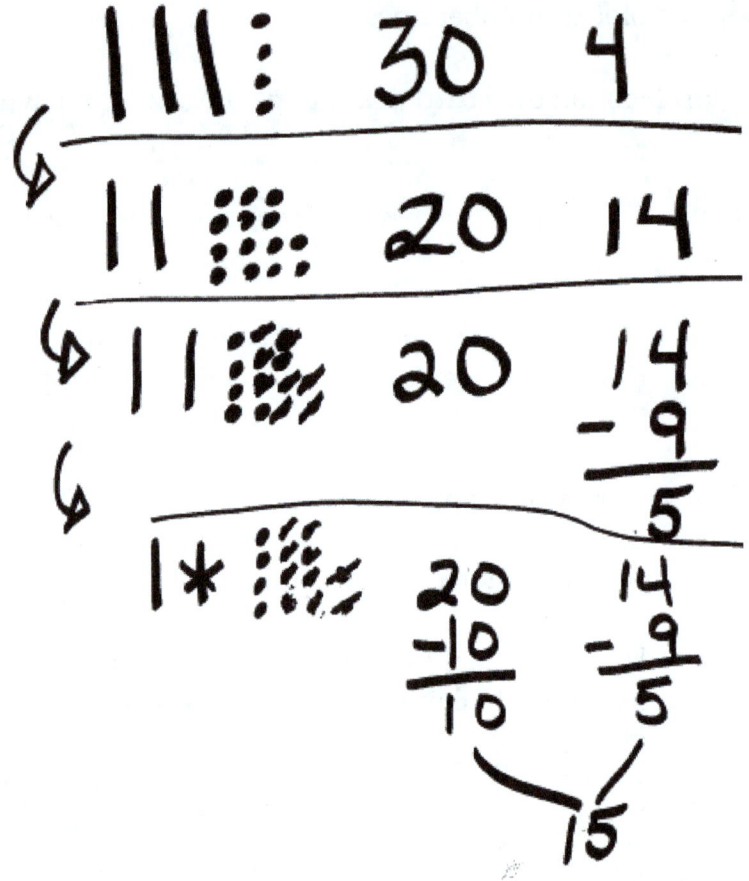

Class Notes:

Separate Start Unknown

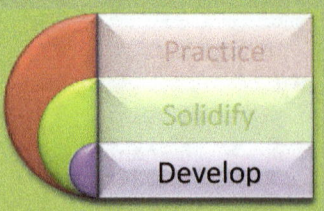

Day: 13	**Objective:** In the context of solving a separate start unknown problem, students will use multiple representations to make sense of it. Students will also begin to generate numerical recordings for addition with regrouping.
Materials:	Unifix™ cubes, base ten blocks, number lines
Launch/ Explore1:	Pose the following problem on the board (chart paper or interactive board) *When the birthday party started, there were some party hats on the table. 8 Children came and they each took a hat to wear during the party. Now there are 14. How many party hats were on the table to start?* Again, allow students to work with their partner to solve this more challenging problem. Encourage them to directly model it and draw a representation that is clear and to represent their way numerically.
Discuss1:	Focus on what is it that they are looking for. What number sentences would match? Why would it make sense for the answer to be a larger number than 14.
Launch/ Explore2:	Pose the following problem on the board (chart paper or interactive board) *When the birthday party started, there were some party hats on the table. 27 Children came and they each took a hat to wear at the party. Now there are 29. How many party hats were on the table to start?*
Discuss 2	Again focus on what they are looking for. Have students generate number sentences to match. If someone used a trial and error approach, highlight and talk about that and why they had to do it (because he was using the subtraction sentence that matches the story).
Exit Slip	None

Excerpt from the classroom:

This problem, more than the others, was the one that tripped up students the most. It was helpful to start out by having students close their eyes as I set up the problem so they could visualize what was happening. Even so, there were still students who thought that there were 14 hats on the table to start. When summarizing the problem, the focus was on the circle model (below) to figure out which part was the unknown.

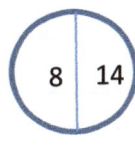

In the second problem, due to the larger numbers, a couple students still had trouble realizing that they should be adding the two figures rather than subtracting. I had to help them act it out so they could understand what was happening, but not as much as in the original problem. Some strategies for solving this ranged from trial and error to using a making a ten strategy. This is consistent with Thomas Carpenter's research in his book *Cognitive Guided Instruction* for separate start unknown problems.

During the share out, we looked at the different number sentences that would match: ____ -27 = 29 and 27 + 29 = ____. Students noticed that it was like a fact family, so we talked about the term "inverse operation" and discuss how addition can be used to solve for subtraction.

Class Notes:

Separate Start Unknown
Trying on another student's thinking

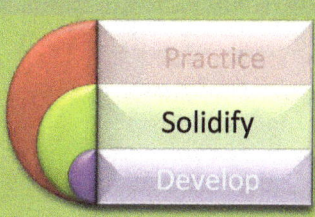

Day: 14	**Objective:** Students will learn to pay attention to another student's strategy and try it on for themselves.
Materials:	Journals, pencils
Launch/ Explore1:	Find a strategy that a student used in the previous problem, such as making tens, and help other students not only make sense of the strategy, but discuss different numerical representations that would match. Have children try out that student's strategy with a similar problem, such as 38 + 17. The point is not doing it the way they want, but to try on someone else's thinking.
Discuss1:	Students have to share with each other and prove that they were using the chosen strategy. Have students share out if it was easy or hard to try on someone else's thinking. Ask them who would be likely to try that thinking on in the future.
Launch/ Explore2:	Pick a second strategy used in solving the problem, such as adding tens then ones (partial sums). Display the strategy and have students discuss what that student was doing and connect it to the blocks, if appropriate. Pose a couple of alternatives for recording that strategy numerically, in case a different way of recording makes more sense to some students. Pose a different problem and have them try on this new way to solve the problem.
Discuss 2	Students have to share with each other and prove that they were using the chosen strategy. Have students share out if it was easy or hard to try on someone else's thinking. Ask them who would be likely to try that thinking on in the future. Make sure the two strategies with different representations are clearly charted for future reference.
Exit Slip	Students hand in journals to make sure that they were able to make a numerical representation for 38 + 17.

Excerpt from the classroom

Today's lesson started by focusing on one student's strategy of making a ten from the problem in the previous lesson. He took one from the 27 to make 30 and then added it to the 26 to get 56.

Using virtual manipulatives found on the website: http://www.glencoe.com/sites/common_assets/mathematics/ebook_assets/vmf/VMF-Interface.html the student showed his friends what he did and as a class we decided how to record that with numbers. Three recording strategies were posed:

Students discussed with their partner about why each matched the strategy of making a ten. After all three were discussed, students were to identify which numerical recording strategy they resonated with more and to apply it to an additional problem of 38 + 17 trying on the "making a ten" strategy".

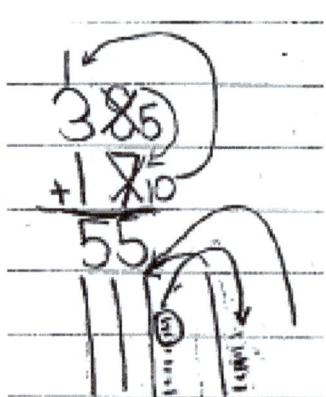

When they were done, the children shared with their partner how they solved the problem and they had to decide if they actually used the "making a ten" strategy or not and how they knew. After discussing, many students said that they liked this strategy and would likely try it on in the future.

A second strategy that was highlighted was one that another student did which was adding tens and ones and recorded by "stringing".

After talking to their partners about what they thought the student did, we explored the possibility of recording it vertically rather than horizontally and how they were similar. I posed an additional problem which they were to try on the strategy of adding tens and ones and to choose one of the two recording strategies. Some students were stuck on the idea of making a ten, thus had a hard time implementing this strategy, more formally known as "partial sums."

Class Notes:

Continue to Try on Thinking

Day: 15	**Objective:** Students will continue to practice trying on another student's strategy and making sense of it for themselves. They will also monitor which strategy makes most sense and is easiest for them to use.
Materials:	Journals, pencils, co-created strategy chart.
Launch/ Explore1:	Refer to the chart created with the two strategies and different notations from the previous lesson and ask students to explain the different representations. Provide students with the handout of four addition problems. Instruct students to solve the first two problems using one or more of the representations on the chart. Confer with different students to either help them make sense of a strategy, or to question them about why the strategy works.
Discuss1:	After there has been enough time for students to finish both problems, allow students to have some time to discuss in partnerships how they solved them and to prove what strategy they used. They should be using the names given to these strategies, such as "making tens," "adding tens and ones," "stringing," "partial sums." Have a couple of students share out their strategy. As a class discuss which ones they chose and which they prefer and why.
Launch/ Explore2:	Instruct the students to solve the third problem on the handout, again using one of the charted strategies, and the fourth problem, using any way they want. Emphasize that it is important that everyone use procedures that make sense, so whatever they choose, they should be able to explain. You will be able to see in the fourth problem if students are taking ownership of any of the shared strategies or are relying on one that they have devised themselves. If any are using a different procedure, have that student explain his thinking. If he is simply using a standard procedure without understanding, then encourage him to either make sense of it or to use one that makes sense.
Discuss 2	After some partner share time, ask students if they chose the same strategies, or different ones, which ones they like the best and why. Pose an incorrect strategy, such as the one in the "Excerpt" below and ask what a student might be doing in this problem and if it is correct or not. Save new strategies for the next lesson.
Exit Slip	Collect the hand-outs for evidence and review.

Let's Try it on!

```
  36          29
 +25         +46
 ───         ───
```

```
  16          52
 +37         +48
 ───         ───
```

Excerpt from the classroom:

We are in the phase of our learning where, for the most part, we are not dependent on the blocks, or pictures, but the numbers seem to be making sense by themselves. The children are able to make connections to the pictures when necessary. It is evident when a child needs a little more time working around the visual/concrete models when he starts randomly doing things with numbers. For example, one student did this:

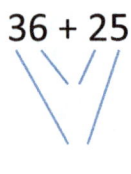

36 + 25

8 + 8

He was confused and was trying to make sense of the example on the chart, but it was necessary to bring out the block and ask him why in the example problem (27 + 29) did the student want to put the "2" and "2" together (since he was obviously thinking of this in isolated digits). He couldn't answer the question, so I built it for him and put the tens together asking him, "why would the student want to do this?" Surprisingly, he had a hard time answering this until he all of a sudden said that they were both tens. Once he made that breakthrough, he was able to see that the tens had to be together and the ones did as well. Now, the numbers made more sense to him and he was able to go back to think numerically.

When looking over the students' work, everyone was able to try on thinking, some trying every strategy. It was obvious from the conversation that students were now taking ownership of one or more ways.

Class Notes:

Continue to Try on Thinking

Day: 16	**Objective:** Students will continue to practice trying on another student's strategy and making sense of it for themselves. They will also monitor which strategy makes most sense and is easiest for them to use.
Materials:	Journals, pencils, co-created strategy chart.
Launch/ Explore1:	Select another strategy that you have observed that is simple and efficient, write it on the board/chart and ask students to privately think about what the student was trying to do. After children have had enough time to make sense of the problem, have them discuss their thinking with a partner, then share out. Ask them to try on that thinking on one of the problems you provide.
Discuss1:	Have a student share how they used the other person's strategy to solve their own problem. Discuss any misunderstandings about the recording of the problems.
Launch/ Explore2:	Do the same for a 4th strategy. By now, someone is bound to be using the standard algorithm. Pose and ask students to try and make sense of the strategy. Have students partner talk and try it on with the second problem on the handout.
Discuss 2	Have students share how they solved the problem with this strategy (standard algorithm). Discuss any misunderstandings about the recording of the problems.
Exit Slip	For the last two problems, allow students to try on any of the 4 strategies discussed. Observe which strategies are being chosen.

Excerpt from the classroom:

In our classroom, the first strategy that was shared today was:

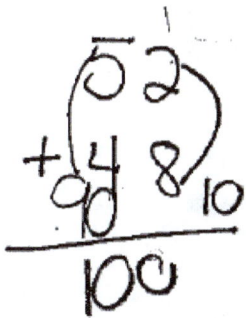

This was like one of the strategies from the first day, adding tens and ones, but recording them off to the side and then putting the total at the bottom.

The second strategy was a student who used the standard algorithm. This was a lot like the way that one student solved it yesterday when he was making a ten and then placing the ten with the other tens. When presented, this strategy initially baffled many students, but as they thought about it and discussed it with their partners, they were able to see that the one on the top was a ten that was made from the 6 and 7. I overheard one student explain to her partner that this was the Japanese way and explained the procedure. She had seen this method used in her Japanese home, but it was interesting that she had not yet associated it to the typical way in America.

Let's Try it on!

```
  55          38
 +26         +43
 ———         ———
```

```
  72          58
 +19         +43
 ———         ———
```

Class Notes:

Using Graphs
To generate questions

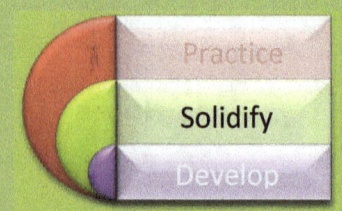

Day: 17	**Objective:** In the context of using a bar graph, students will generate and answer questions from the data
Materials:	Journals, pencils, chart with a bar graph with data.
Launch 1	Show the class the bar graph and ask students what it is all about. Discuss the interval chosen. In this case, it is an interval of 3 and ask why they think I chose to do that rather than 1's. Ask students to talk with their partner to see what kinds of questions they can ask using the data. At first, they might ask questions like, "does Mr. Dugan's class like to read books?" Tell them that you cannot completely answer a question like that with the data, although we might be able to infer an answer. Ask them to come up with some questions that can be answered with the data provided.
Explore/ Discuss 1:	Ask for a suggestion, such as one that is data related (i.e., median, mode) or one that would use addition to solve, and write the question on a presentation board or chart for children to solve. When adding, refer back to specific strategies from the previous days.
Launch 2	Pose a question that requires students to compare two of the data points, such as: *How many more books did Mr. Dugan's class read than Ms. Cleveland's class?*
Explore	Students work with their partner to try to come up with the answer to the problem. It is very likely that they will add the two pieces of data rather than find the difference because they are not fully understanding the question. Since they are working together, notice if the partnerships are listening to each other and agreeing on an answer, or are they still working in side-by-side play.
Discuss	Share strategies for this initial compare problem. Have the student who added up share first. Some students will say the answer is 2 because the increase on the graph is two spots, but they are not accounting for the fact that each space represents 3 books, rather than 1. Save subtraction strategies for the next lesson.
Exit Slip	none

Excerpt from the classroom:

In today's lesson, I must admit that when I asked students to come up with questions related to the graph, I wasn't expecting the kinds of questions I got. Some partnerships were looking for patterns in the numbers on the left side of the graph (interval). Others were asking questions as to whether or not Ms. Patrick's class liked to read or not because her class had a lower number of books. We had to talk about how graphs should make us wonder such things, but that some of those questions we could not answer because we did not know all of the facts. For example, Maybe Ms. Patrick's class read longer books, therefore could not read as many. I had to provide examples of questions that could be specifically answered by the data using numbers.

Also, when looking at the difference between the number of books that the two teachers read, I was not expecting a student to think that it was only 2 more until I realized that he was looking at the number of squares rather than the numerical difference. Another student said 3, but she did the same thing, only she did not start counting on beginning with the next number.

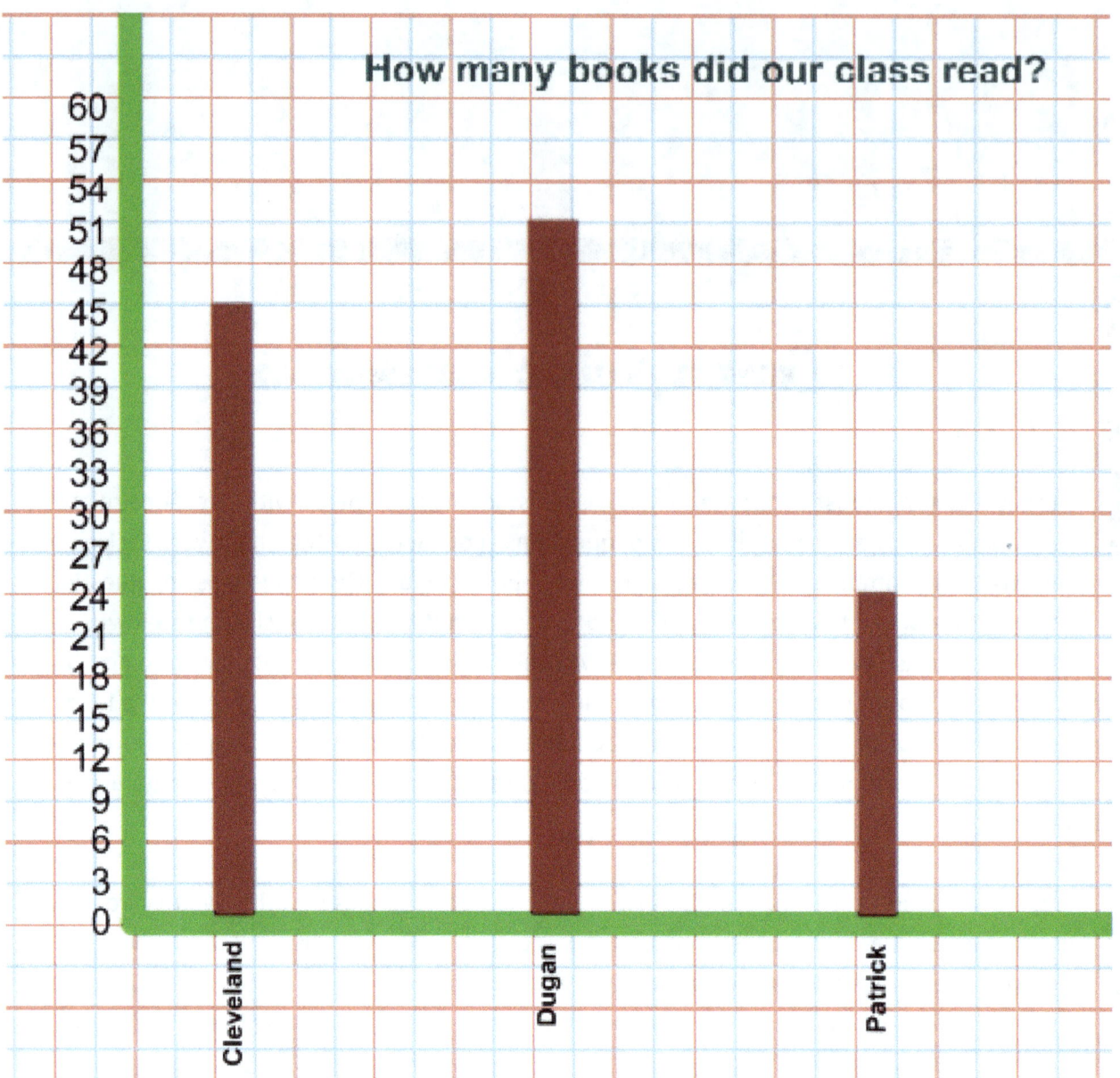

Class Notes:

Using Graphs
To compare data

Day: 18	**Objective:** In the context of using data from a bar graph, students will explore subtraction strategies to compare two data points and continue to explore how to numerically record subtraction with regrouping.
Materials:	Journals, pencils, chart with a bar graph with data, base ten blocks.
Launch 1	Show the class the bar graph from the previous lesson and ask students what they remember about it. Remind students from the charted strategies the ways that students had solved the problem of how many more books Mr. Dugan's class read than Ms. Cleveland's. Mention that in addition to the strategies shared, you saw some students solving the problem with subtraction. Have students talk with their partners as to why they would choose to do this. After some partner talk time, explore the notion of inverse operations using smaller numbers, for example: 3 +____ = 8 so 8 – 3= _____. In this problem, many students added up from 45 to 51: 45 + ____ = 51, therefore, the problem can also be solved using 51 – 45 = _____.
Explore 1	Have students work with their partners to find a way to subtract 51-45. They already know the answer will be 6, so the emphasis is on how to solve the problem.
Discuss	At this point, most students will most likely still be either using base ten blocks or drawing tens and ones in their journals to solve the problem. If someone used a strategy with just the numbers, then have him share first and perhaps prove his strategy with base ten blocks. Then go back and walk through the problem step by step asking the class how each step could be recorded numerically.
Explore 2	Pose the following problem and ask students, with their partners, to pick one of the strategies that were charted to solve: 42 – 37. These numbers are not on the graph, but are similar to the previous problem so that students can more readily apply one of the strategies that surfaced.
Discuss	Select students to share their strategy that matched one of the ones charted.
Exit Slip	Monitor journals for strategies chosen to try on and accuracy.

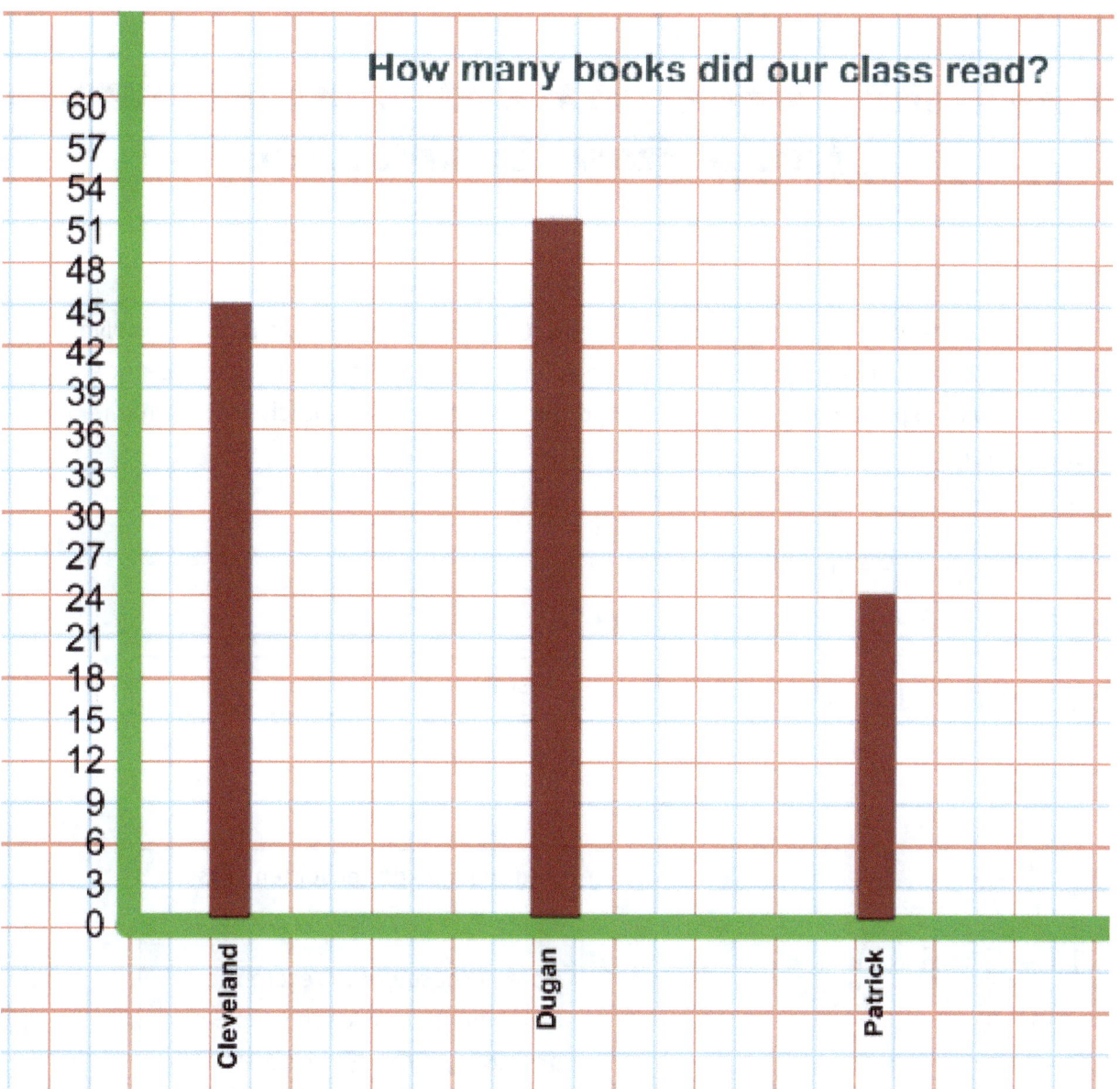

Excerpt from the classroom:

In our class, most every student chose to get base ten blocks or to draw a picture to solve, however, they are stuck on the numerical recording for the subtraction. We had a student go step by step to explain what she drew and had the rest of the class match it with a numerical representation:

51 – 40 = 11

First she took away 4 tens

11 – 5 = 6

Next she traded a ten for ten ones and took away 5 ones

Note: I chose not to record the trade numerically at this time, as I am saving it for a future lesson. I am focusing on the 11-5 at this time.

Class Notes:

Using Graphs to Compare
Trying on another student's thinking

Day: 19	**Objective:** In the context of trying on another student's strategy, children will practice subtracting tens then ones.
Materials:	Journals, pencils, chart with a bar graph with data.
Launch	Of all the strategies that have surfaced, review the strategy of subtracting tens then ones using the chart it was recorded on (note: even though the student may have traded in her picture, we are not going to focus on the recording of that quite yet). Tell students that today they are going to try on this strategy. Have the students give it a name, like subtracting tens then ones, or name it after the student who suggested it. Pose the problem, 34-28 and ask them to solve it using the said strategy.
Explore	As students are trying to solve this problem, many will not be using the strategy of subtracting tens then ones. Some will still be just drawing, and others will revert to their own way. Remind all students the importance of learning to analyze and try on someone else's thinking. If students choose to solve the problem on their own at first, make sure they have had time to discuss with their partner how their way matches that of the one on the board.
Discuss	Have different students share a component of how to solve the problem. For example, one student might tell you how many tens and ones to draw, another can tell how many tens to remove, another can state how to record that part in numbers, etc.
Explore 2	Have students continue to solve problems using the same strategy. After about one or two more problems with drawing and recording with numbers, have students try to solve the problems with just numbers. Use the following problems: 64 – 57; 45 – 36; 82 – 77; 51 – 48, since after subtracting the tens, there will be one ten and some ones left over.
Discuss	Continue to share out and start to focus more on the numbers and less on the picture, depending on the readiness of the students.
Exit Slip	Monitor their math journals to see if they are trying on the strategies.

Excerpt from the classroom:

Even though we had a successful week of trying on other students' addition strategies, several children really struggled when trying to make sense of this subtraction one. I chose this strategy because it mirrored a lot of one of the addition strategies, adding tens and ones. I also chose not to focus on recording the trade as it would take away from the simplicity of this strategy, even though that is what they had to do with their pictures in order to subtract. In a couple of days, once they are able to really make sense of what people are doing with the numbers, we will focus on figuring out how to record regrouping strategies.

Class Notes:

More Comparing with graphs

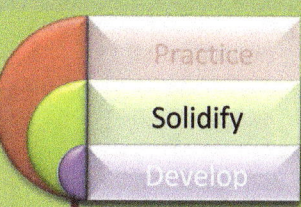

Day: 20	**Objective:** In the context of using data from a bar graph, students will solve another comparing problem that involves regrouping using any strategy they choose.
Materials:	Journals, pencils, chart with a bar graph with data, base ten blocks.
Launch	Review the strategy from the previous day that was highlighted, subtracting tens then ones. This strategy should be on a chart that is visible for all to see. Go back to the graph of how many books each class read. Tell them that in the past two days they were finding out how many more books Mr. Dugan's class read than Ms. Cleveland's class. Today they need to find out how many more books Mr. Dugan's class read than Ms. Patrick's class. They may want to try on the charted strategy that they practiced in the previous lesson, or try on one of their own. Reinforce that if they use a picture, they need to make write in numbers what they build/draw.
Explore	Students should be working with their partner to come up with ways to solve this problem. They may choose to work privately at first then compare their solution strategies. When going around to the different partnerships, ask them how they solved the problem and ask students how their partner solved it, to hold them accountable for listening and making sense of their partner's work. Notice who took on the charted strategy of subtracting tens then ones, who are using number line models, who are using base ten blocks, and who are using other numerical recording systems.
Discuss	Start the discussion by having a student share how they used the previous strategy. In this case it will be 51 – 20 = 31; 31 – 4 = 37. Have students share out a couple of different number line strategies, and then share out any other simple numerical recording. Save the base ten block strategy for the next lesson, since this will be a lesson all to itself.
Exit Slip	Review the journals to see the different systems used, who are still using a counting up strategy, who are building but not recording, who are able to build and match with numbers, and who are able to use efficient numerical representations.

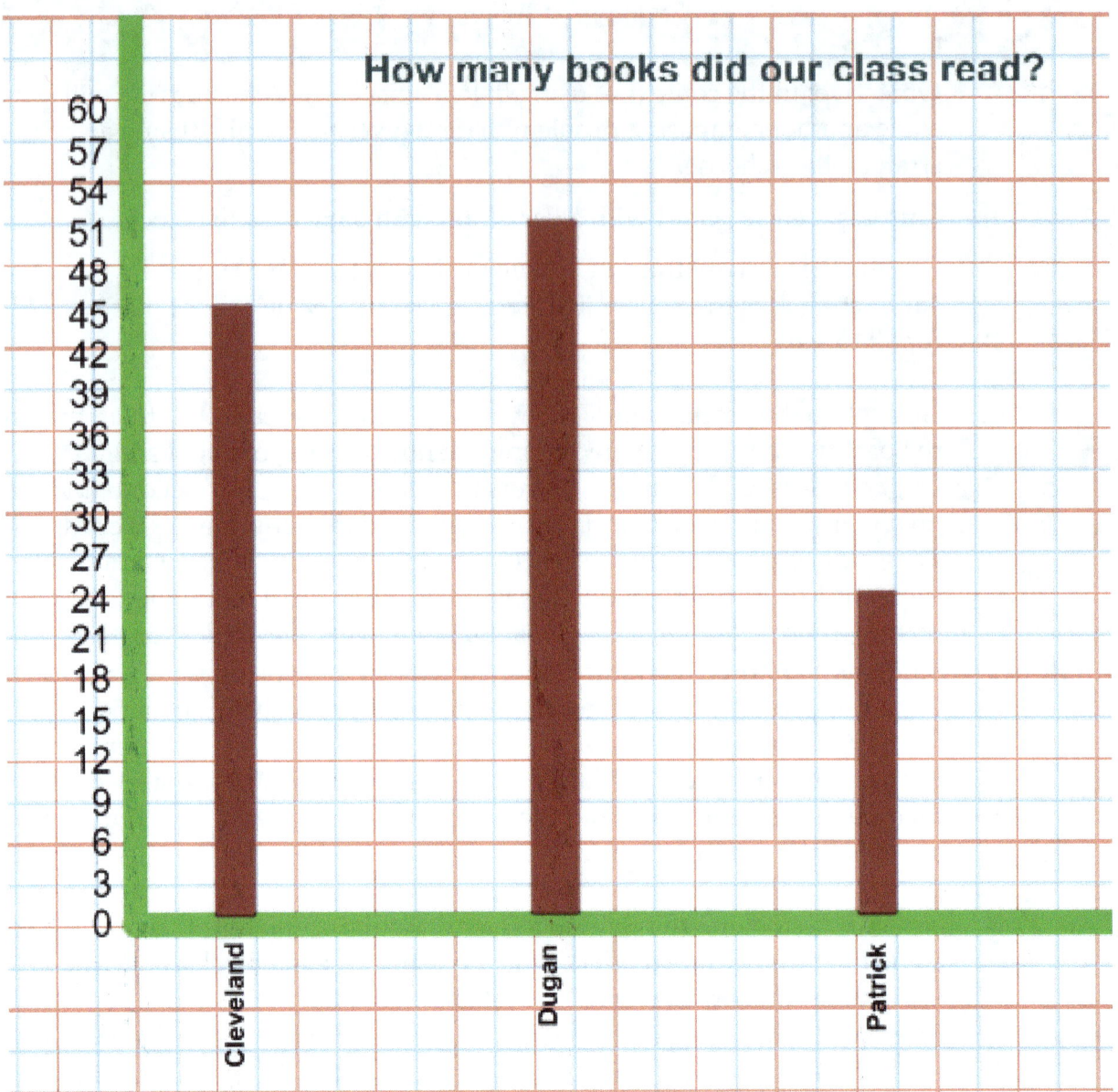

Excerpt from the classroom:

What I noticed from today's lesson was that about 1/4 of the students tried on the subtracting tens and ones method from yesterday. Another ¼ tried using some kind of number line to either add up or subtract. Below were two of the ways that students used number lines that were shared during the discussion. The top one was an "adding on to" model, and the bottom strategy was a student who started at 51 and was subtracting back until she used up 24. In both cases students had to know what to keep track of and where their answer was.

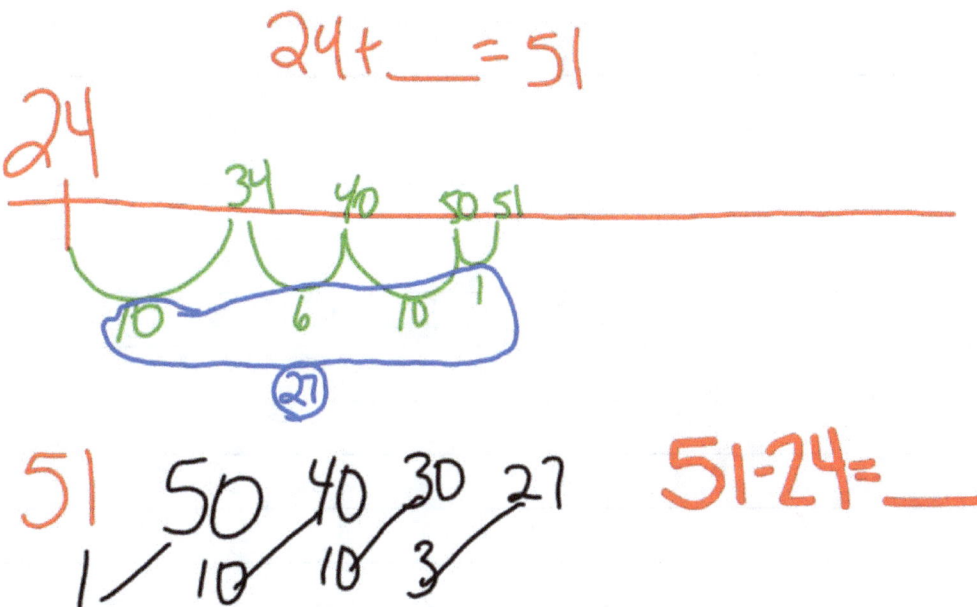

The other half of the students went to get base ten blocks or just drew the picture of blocks. Most started by first removing the tens, trading a ten for ones, then removing the remaing ones. What I thought was interesting was none of these students used numbers to record their strategy. This told me a lot about where they still were developmentally in this process, so I am going to be careful on how I structure's tomorrow's lesson. Since students are using number lines as a way to reinforce counting strategies, I am choosing to not chart these on our permanent strategy chart. I want to reserve the chart for strategies that lead to a deeper understanding of number and place value.

Class Notes:

A Look at Renaming

Day: 21	**Objective:** In the context of using expanded notation, students will practice the numerical recording of renaming numbers when trading.
Materials:	Journals, pencils, chart with a bar graph with data.
Launch	When starting this lesson, remind students of the problem they were solving from the previous lesson. Briefly review the strategies that were tried, including the ones who used subtracting tens and ones, the number line strategies, and building/drawing a model. Tell them that it was noticed that many students used base ten blocks or a picture of them, but none of them used numbers to describe what they did (if that is the case in your classroom).
Explore/ discuss	On the interactive white board, show step 1 (on the following page). Ask students to explain what this means. Have them copy the picture and the numbers that they see on the board. Then show step 2. Have the children partner talk to discuss what is happening in this picture, then share out and have them copy the cross outs and the numbers into their journal. Show step 3. Again, students partner talk to make sense of what has happened in the problem, share out and discuss why it was necessary to trade the ten. Have them pay particular attention to the numbers and why they changed from 30 + 1 to 20 + 11. Emphasis in this part of the discussion will be around the fact that the amount is the same, but it has been renamed. Make sure students have copied the changes. Have the class predict what will come next and then show step 4. Again, discuss with a partner, then as a class. Show the next slide of the alternative recording of the same problem, have them partner talk to discuss if it is the same or different to what they just explored and how. Allow children to draw as needed to illustrate their thoughts.
practice	The one component to regrouping that is difficult is renaming the numbers, so show the slide with 45, 83 & 24 written on it. Ask students how to write 45 in expanded notation. When they tell you, write/type it in below (40 + 5). Let them know that this is its name. Tell them that they are going to trade the ten and rename it. Ask students what its new name will be. If students are confused, add the model next to the numbers and break up one ten to ones and move them to the ones side. Write the new name (30 + 15 below the arrow). Do the first 3 together as needed and have the children do the bottom 3 alone as an exit slip, individually conferencing with each student until all understand what to do.
Exit Slip	Expanded notation form.

Step 1

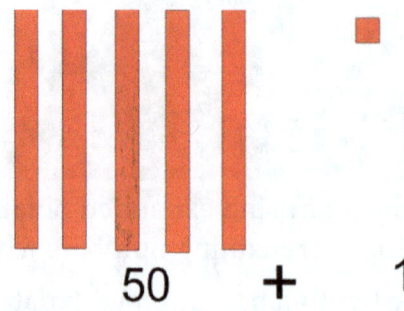

Step 2

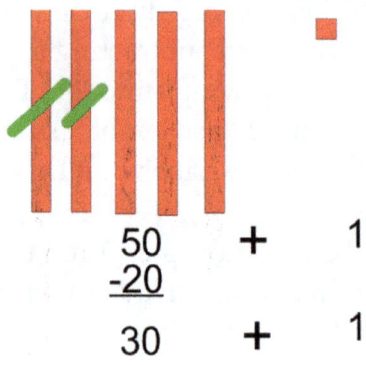

Step 3

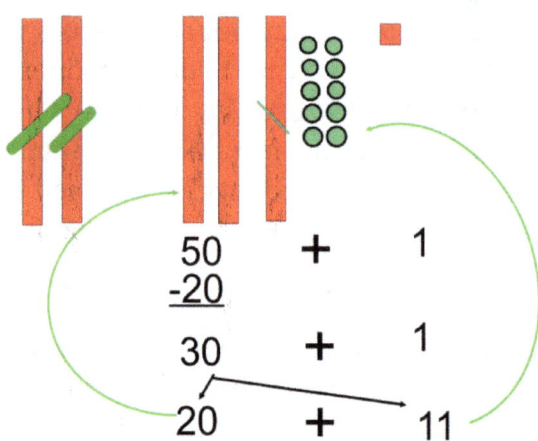

Step 4

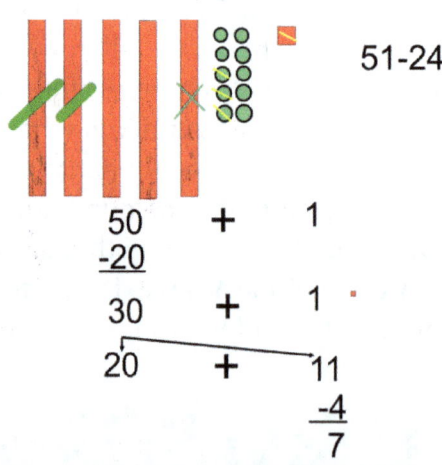

51-24

84

Alternative Recording

$$\begin{array}{cc} 40 & +11 \\ \cancel{50} & +\cancel{1}\;\blacksquare \\ -20 & 4 \\ \hline 20 & 7 \end{array}$$

Write each using expanded notation

45 ∧ ↓

83 ∧ ↓

24 ∧ ↓

32 ∧ ↓

63 ∧ ↓

51 ∧ ↓

Excerpt from the classroom

Because the students were having a hard time recording their pictures with numbers, this lesson was more guided that the others so they could see how to match the numbers with the models they were already building/drawing. Also, knowing the difficulties around the recording of traditional subtration with regrouping algorithms, I know how important it is to make sure children really understand what is happening conceptually with both the model and numbers. Therefore, I have chosen to use an expanded notation model so that the place value of the digits is preserved when subtracting. Otherwise, there is a huge risk for students to subtract one from the tens and literally add one to the ones. In fact, today on the expanded notation worksheet, a couple of students did just that. For example, they wrote 51 as 50 + 1, then renamed it to 49 + 2. It is important that these students continue drawing when using numbers because they are loosing meaning of them.

A cute thing that came out of today's lesson was that one student said what we were doing when renaming the number was giving it a "nickname". Therefore, the class decided to call this strategy the "nickname" strategy.

This strategy will be recored with the subtracting tens and ones strategy and reviewed in the next lesson so students can practice trying on one of the two. I plan to maintain the expanded form for a couple of days until students begin to take ownership of it. After that, we will look at students who are using the traditional algorithm, because there are one or two who are already doing that on their own.

Class Notes:

Trying the Renaming Strategy

Day: 22	**Objective:** Students will try on the renaming strategy, "a.k.a. nickname strategy" as a way to make sense of the numerical recording of what they have been doing with the base ten blocks.
Materials:	Journals, pencils, base ten blocks available as needed
Launch 1	Review the two subtraction strategies that are now on the permanent subtraction strategy chart: subtracting tens and ones, and "renaming" or the "nickname" strategy. Ask students to try both of these strategies while solving 34 – 16.
Explore/ discuss	With a partner, students first solve it alone then share out which strategy he/she used and how they know. Share out as a class and decide which one was easier. The "renaming" strategy will be more difficult for the group because it is newer, but it is a very important strategy, not only because it leads toward understanding of the traditional algorithm, but because it is recording numerically what students have been doing with the base ten blocks.
Launch	Pose a new problem, 57 – 29. Ask students to try on the "renaming strategy" again.
Explore/ discuss	Confer with students/partnerships as they solve the problem to identify trouble areas. Encourage students to use base ten blocks if they are getting lost in the numbers. At this point, all should be very comfortable with subtracting using the blocks, so help them record what they are doing.
Exit Slip	Have them write the number 46 in expanded form and "rename" or "nickname" it (see problem below). This will help gauge the readiness of students to use this strategy without a picture/model. Meet with students privately or in a small group if they are still stuck with this concept.

Write in expanded form and "Nickname" the number below:

Excerpt from the classroom:

As a class, the children were quite comfortable around the subtraction of tens and ones strategy, however, they struggled a lot with the renaming strategy. This is because they were too focused on the numbers and not the model. No one elected to get base ten blocks for the first problem of 34 – 16, however, many did try to draw a picture. By the second problem, several students ended up being encouraged to use the blocks because they were getting lost in the numbers. They were trying to follow the strategy that was written on the interactive white board from a previous problem. The issue for most students was not in the renaming of the numbers, but in what it was they were suppose to subtract. Most students were able to rewrite the 30 + 4 to 20 + 14, but then started taking away random numbers rather than the 10 and the 6. Another error was when students did not rename the number and did the classic "4 – 6 = 2". Both of these errors were indicators that students were not making the connection to the models they had been building/drawing previously, or in other words, were dropping the models too quickly. I spent the explore time meeting with each partnership to help make the connnection more explicit for those who were working more abstractly than they were ready. I am anticipating it taking just one more lesson for the majority of students to take more of an ownership of this strategy.

Subtracting tens then ones

Rewriting in expanded notation then renaming to subtract.

Example of not "renaming" and subtracting incorrectly

Class Notes:

Further Exploration of the Renaming Strategy

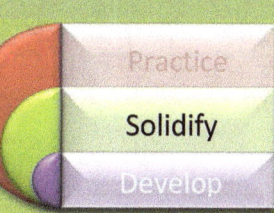

Day: 23	**Objective:** Students will continue to try on the renaming strategy, "a.k.a. nickname strategy" as a way to make sense of the numerical recording of what they have been doing with the base ten blocks.
Materials:	Journals, pencils, base ten blocks available as needed, co-created strategy chart
Launch	Review the "renaming" or the "nickname" strategy. Provide a handout with four subtraction problems and instruct them to work on the first problem with their partner.
explore / discuss	With a partner, students first solve it alone then discuss. Each need to be attending to the other to ensure understanding. The teacher will be conferring with the different partnerships to monitor understanding by asking them to defend their thinking, perhaps connecting it to a picture or model. Due to the complexity of this strategy, students will be working at different paces. Allow students to practice this same strategy on the remaining problems. When they have finished the 4 problems on the handout, they can make up their own in their math journal. End the lesson by discussing the improvements and "lightbulbs" you saw during this activity and, although it may be difficult, the importance of learning to try on someone else's thinking and making sense of the numbers.
Exit Slip	Students turn in their hand out as evidence of their work.

Let's Try it on!

43 – 17 **34 – 18**

65 – 36 **51 - 36**

Excerpt from the classroom:

They say Learning is messy, and today was definitely one of those days. Children do not learn at the same rate and it is imperative the teachers know what the developmental sequence is for learning, so they can recognize exactly where their students are and what is that next step the teacher should be pushing the child towards. In this and the previous lesson, children were asked to try on a strategy that they have been having trouble recording numerically, but have been quite successful in building/drawing. In this lesson, some students were definitely in their practice phase of learning, and were able to practice using this strategy in a more abstract way. Other students were struggling and it was important to allow them to get the base ten blocks and have them build it. As they built each step, I would ask them what they thought they should do with the numbers, or I might suggest by writing what they did. Other students were somewhere in the middle. They were able to depend on their own drawings to make sense of the problem and used the strategy example on our chart to apply to their own learning. Since there was a large discrepancy between the levels of understanding, I allowed the students to keep practicing more problems. Today's session was full of lightbulb moments. They are not only learning to make sense of this strategy, that will eventually lead to the standard algorighm, bot they are learning to take something that is difficult and make sense of it. It was important that several problems of similar type were available because with each problem, the ease of the computation increased.

Example of a student who needed to make the connections for sense-making.

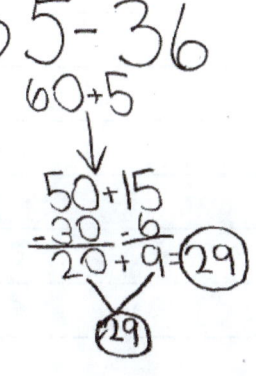

Example of a student who was able to solve in a more abstract manner.

Class Notes:

More Data to Analyze

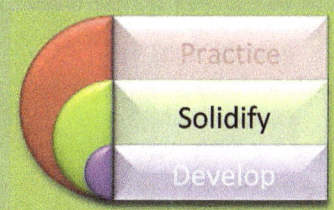

Day: 24	**Objective:** In the context of analyzing data on a double bar graph and answering questions, students will try on one of the addition or subtraction strategies that have been charted from the previous lessons.
Materials:	Journals, pencils, chart with a bar graph with data (or on interactive white board), and strategy charts.
Launch 1	Review the different addition and subtraction strategies that are now on the strategy charts hung up in the classroom. Show the slide/chart of the bar graph with the additional set of data. Ask students to brainstorm what kind of questions they could ask that can be answered by the data on the chart and what kind of questions they are wondering that cannot be answered by the data. After a few minutes of private think time and partner talk time, chart their questions.
Explore 1	Once there is a fair selection of questions, allow partnerships to choose one to answer. Remind students that no matter what they chose, to try on one of the strategies that have been explored as a class and are on the charts. Make sure that the charts are displayed. Go around and confer with the different partnerships regarding the question they chose and the strategy they used. If partners solve using a number line strategy, encourage them to also solve using one of the ones charted.
Discuss 1	After groups have had enough time to pick a question and solve it, have a few share out what they chose to find out. Ask questions such as how they knew what to do to solve their problem and what strategy they used to solve it.
Launch/ explore 2	Give feedback, such as how they are working as partners and if they are selecting questions that can be answered by the data. Allow students to continue working as partnerships to select further questions to solve.
Discuss	As you are conferring with the students, notice which strategies they are using. Select 3-4 students to write their strategies on a chart paper that can be used during the discussion. Have those students tell their peers what question they were trying to solve and how they solved it. Ask the students to identify what strategy the students used.
Exit Slip	Monitor journals to see if they are solving the question they chose, and what strategy they are using.

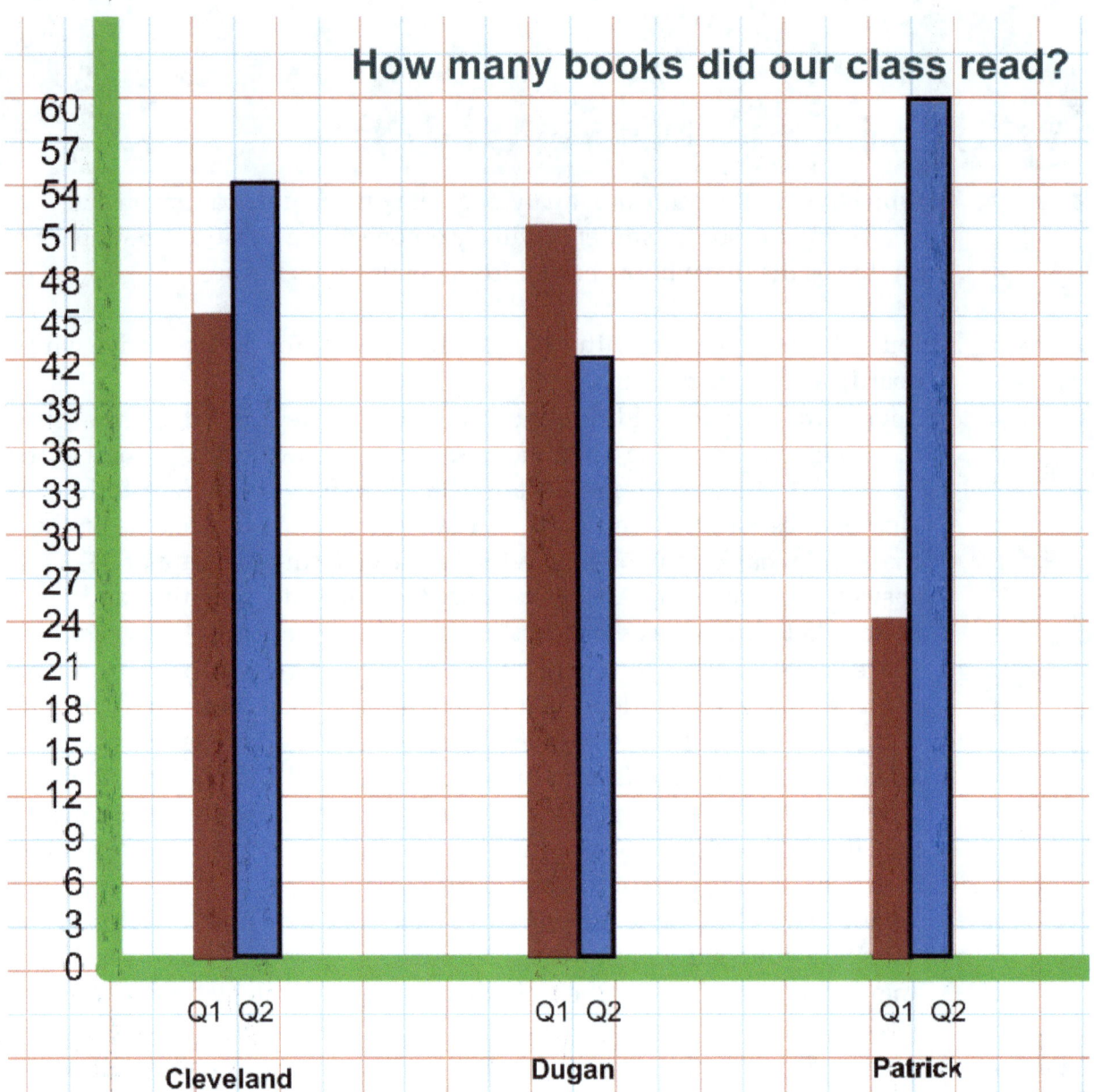

Excerpt from the classroom:

Since the original bar graph had been used for a few days, showing them this graph with the added data did not throw the students. They knew exactly what the extra data represented. Since I just wrote "Q1" or "Q2" on the bottom of the columns, many students didn't exactly know that it meant "quarter," but all knew that the bar to the left was the number of books that was read before and the bar on the right was the number of books that was later. The exaggeration in the number of books read by Ms. Patrick's class from the first to the second quarter was quite exciting for the class when revealed.

Some questions that the children selected were:

"How many more books did Ms. Patrick read in the 2nd quarter than Mr. Dugan's class in the first quarter?"

"How many books did each teacher read in both quarters 1 and 2?"

"How many more books did Mr. Dugan's class read in quarter 1 than quarter 2?"

"What is the number together if you add all of quarter 2's data?"

"Is there a mode in Quarter 2?"

Some partnerships chose to answer a question they posed that was not shared, and others chose to answer one of the above questions.

When solving the problems, I had some students who initially went straight to the number line model, so they needed to be encouraged to try on one of our charted strategies. It was really exciting to go around to hear about the different questions they were trying to solve using the graph.

> What is the answer if you add all the quarter ones?
> 50+40+20=110
> 10+1+5+4=120

A student who used adding tens and ones strategy.

What I liked most about this lesson was that students were able to select questions that required calculations of several numbers if they wanted. I was actually surprised as to how a few students were able to mentally calculate several two digit number. This proved to be great for differentiation.

Note: This lesson took more than one day.

How many books did all the classes read all together for Q1 and Q2.
How many did Mrs. Cleveland's class read for Q1 and Q2 all together?
How many did Mr. Dugan's class read for Q1 and Q2 all together?
How many did Mrs. Patrick's class read for Q1 and Q2 all together?

$45 + 54 = 99$
$99 + 51 = 150$
$150 + 42 = 192$
$192 + 24 = 216$
$216 + 160 = 276$

Cristol's Way
$34 - 16$
$34 - 10 = 24$ +1:1
$24 - 6 = 18$

NickName
$80 - 16$
$80 + 0$
$-10\ 6$
$60 + 4 = 64$

NickName
$100 - 25$
$100 + 0$
$90 + 10$
$-20\ 5$
$70 + 5 = 75$

Nick Name
$34 - 16$
$-10 + 6$
$10 + 3$
18

NickName
$100 - 85$
$100 + 0$
$90 + 10$
$-80\ 5$
$10 + 5 = 15$

NickName
$85 - 61$
$80 + 5$
$70 + 15$
$-60\ 1$
$10 + 14 = 24$

Class Notes:

Practicing Subtraction

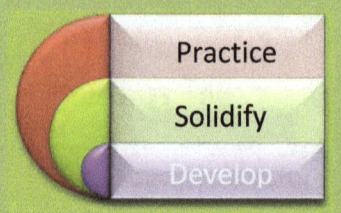

Day: 25	**Objective:** In the context of answering a question regarding the double bar graph, students will try on one of the subtraction strategies that have been charted from the previous lessons.
Materials:	Prompt recording sheet, felt-tip pen, chart with a bar graph with data (or on interactive white board), journals, pencils, and strategy charts.
Launch	Present students with a recording paper with the prompt written on it: "How many more books did Ms. Patrick's class read in Q1 than in Q2? Give a possible reason that her score may have increased so much. Name whatever strategy you chose to use." Read and minimally discuss the prompt and let students know that as partnerships they will solve and have to decide how to record their work using the felt-tip pen. The purpose of the pen is for them to have first practiced what they will write in their journals and negotiate what they will put on their recording sheet, minimizing errors and erasing.
Explore	Students can either solve together with their partner, or separately and then compare strategies. They have to come up with how they are going to represent their thinking on the charts.
Discuss	This discussion will be individualized to the different partnerships as the teacher goes around and confers on what they are doing. Notice things like the differences/similarities between the students' strategies. Ensure that they are answering all the questions and that they are both contributing to the work.
Practice	Provide students with a practice sheet with subtraction problems. Individually check on the students and ensure that they are carrying over their strategies to solve these problems.
Exit Slip	Collect the practice page and monitor which students are still not trying on one of the subtraction strategies and arriving at erroneous answers.

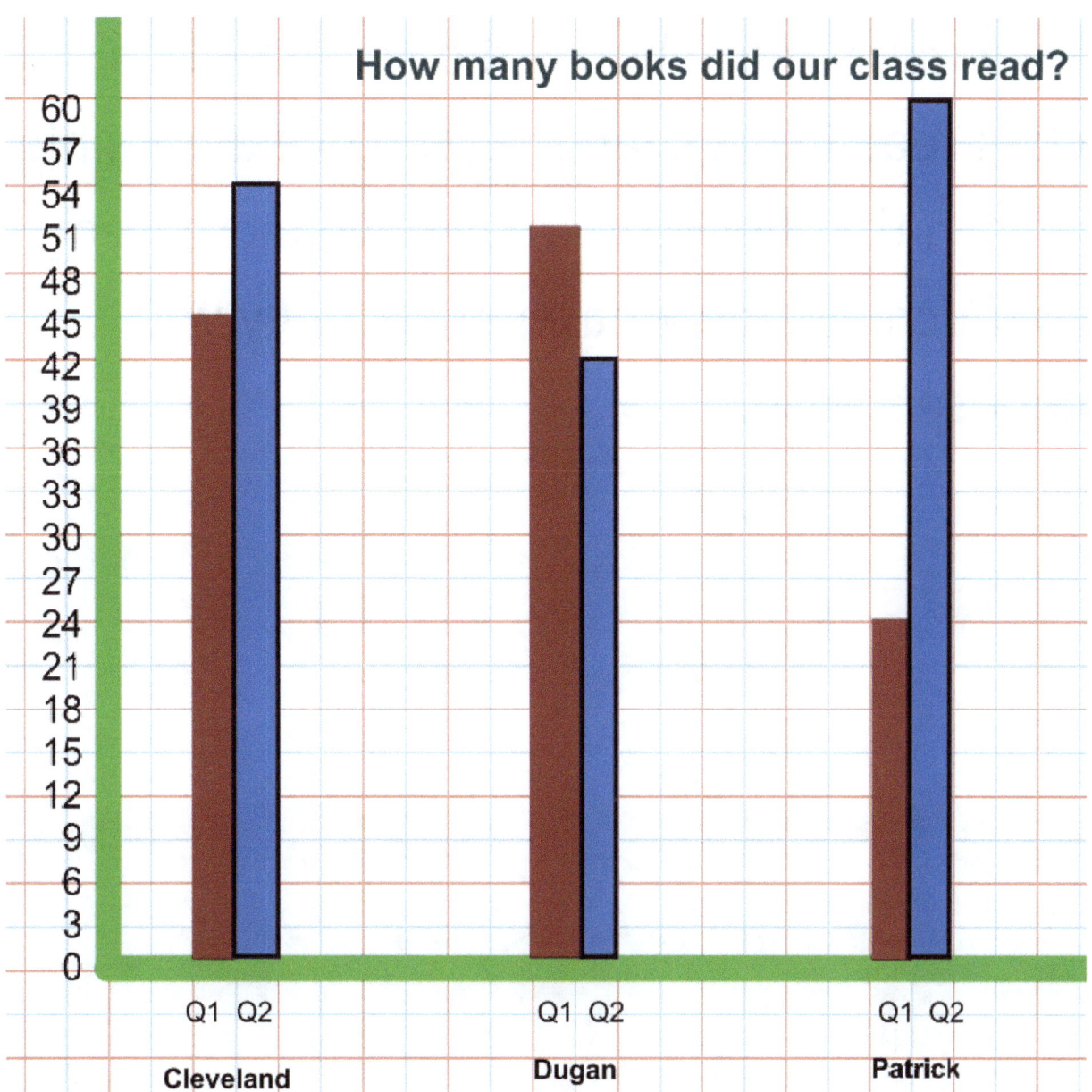

How many more books did Ms. Patrick's class read in Q1 than in Q2? Give a possible reason that her score may have increased so much. Name whatever strategy you choose to use.

25 − 19 =   or   25
                                          -19

68 − 29 =   or   68
                                          -29

45 − 38 =   or   45
                                          -38

Excerpt from the classroom:

This was the students' first experience to create a math poster and using the felt-tip pens. There were some minor management issues involved, such as a couple of students writing on themselves, but overall the children used them well. We did not hand out the pens until the children had already solved the problem in their journals and decided what they were going to write. However, some students showed their impulsivity and as soon as they received the pens wanted to start writing before they planned out what was going to go where.

What was the most interesting about today was that even though the students did not seem to have a problem trying on one of the two subtraction strategies, when they were given the practice sheet that just had "naked numbers" on it, a couple went straight to an erroneous subtraction system and threw out the window everything they had just done on the previous problem. Since those students were loosing the meaning of the numbers as soon as the context was removed, we had to emphasize that they had to show one of the strategies, preferribly the one they had just used and already proved they understood.

Today, one student tried solving 60 – 24 this way:

$$60 - 24$$
$$40 + 4 = 44$$

Her partner knew that it was incorrect because she did not regroup, but this student was not convinced of why her strategy did not work. I drew a picture of a number line and modeled 0-4 and showed her that it would be -4. Without a lot of explanation around negative numbers, they both understood it was not 4 and less than zero. I helped her rewrite her strategy to then look like:

$$60 - 24$$
$$40 - 4 = 36$$

The partner exclaimed, "That makes so much sense!" I plan to use the above misconception tomorrow for the students to analyze, since it is the most common subtraction error, and I wonder if either of these students will contribute to the conversation by using the number line as a mode.

Class Notes:

Practicing Subtraction
By analyzing a mistake

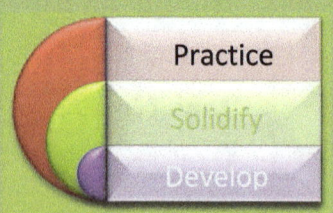

Day: 26	**Objective:** In the context of analyzing an erroneous strategy, students will explore more deeply the need for trading in a subtraction problem.
Materials:	Chart or presentation board with example problem with student's erroneous strategy, journals and pencils
Launch / Explore	Present the problem on the board and say that on the previous day you noticed a couple of students solve the problem this way (even if you did not). Ask students if the strategy works and how they know. Allow them to work as partnerships or 2 partnerships together (groups of 4). They need to figure out what the student did and prove why it does or does not work. Initially, students will most likely be agreeing, but not writing in their journals anything to prove it. Encourage using a drawing, adding back to see if the sum is the same as the total amount they started with, etc.
Discuss	After an ample amount of time for discussion in groups of 2 or 4, ask the students how many of them think that the strategy worked, how many think it does not work, and how many are not sure. Start with someone who is not sure to explain what they are thinking thus far. The main part of the discussion will be around the notion of 0 – 7, so you may want to use the number line as a different way to model this problem, starting at zero and counting back 7 and landing on -7 rather than 7.
Practice	After the discussion, provide some more practice for subtraction and monitor the strategies the students are using.
Exit Slip	Collect the practice page and notice which students are still not trying on one of the subtraction strategies and arriving at erroneous answers.

Does this strategy work? Why or why not?

-discuss with a partner how you can prove it

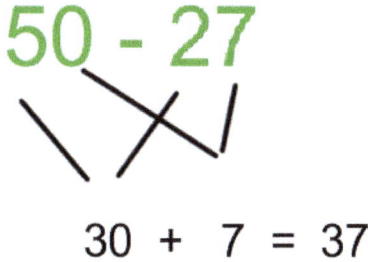

30 + 7 = 37

A number line might help

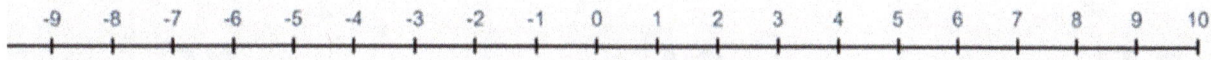

If the conversation uses a number line to get -7, you might want to share this slide:

So really, if you do it this way, it is...

50 - 27

30 - 7 = 33

Name_____

```
  83        91        75
 -67       -22       -27
 ---       ---       ---


  54        25        91
 -18       -16       -76
 ---       ---       ---


  61        53        95
 -17       -19       -48
 ---       ---       ---
```

Excerpt from the classroom:

I am not sure why, but I was surprised that as many students thought that the problem was correct as they did. I think that because most had not thought about subtracting with the "stringing" model, it was pretty suggestive and it seemed ok to them. It was very important for this lesson to give ample amount of discussion time. It would had been very easy to have ended the discussion at just 5 minutes, but at this point, not one student had written anything in their journals to prove whether it did or did not work. The conversation was at a very superficial level at this time. It took about another 5 minutes of students working side-by-side and "sharing" their answers, but students were not really thinking deeply at what their partner was saying. One partnership was kind of eavesdropping on another and they kind of morphed into one group and that is when the conversation deepened. I had the other partnerships team up and that did shift the conversation. Noting that there wasn't a lot of accountable talk happening intially, I also notified them that during the discussion they were not going to talk about their own thinking, but what their partner said. This gave them a reason to listen even harder and make sense of their partner's thinking.

Class Notes:

Practicing Subtraction
Analyzing another mistake

Day: 27	**Objective:** In the context of analyzing an erroneous strategy, students will explore more deeply the need for trading in a subtraction problem.
Materials:	Chart or presentation board with example problem with student's erroneous strategy, journals and pencils
Launch / Explore	Present the problem on the board and say that on the previous day you saw a couple of students solve the problem this way (even if you did not). Ask students if the strategy works and how they know. Allow them to work as partnerships or 2 partnerships together (groups of 4). They need to figure out what the student did, and prove why it does or does not work. Initially, students will most likely be agreeing, but not writing in their journals anything to prove it. Encourage using a drawing, adding back to see if the sum is the same as the total amount they started with, etc.
Discuss	After an ample amount of time for discussion in groups of 2 or 4, ask the group how many of them think that the strategy worked, how many think it does not work, and how many are not sure. Start with someone who is not sure to explain what they are thinking thus far. The main part of the discussion will be around the notion of 0 – 8, so you may want to use the number line as a different way to model this problem, pictures of base ten blocks, or bringing in a strategy discussed to see how it would work with that.
Practice	After the discussion, provide some more practice for subtraction and monitor the strategies the students are using. Make sure that today they are working independently and not with a partner.
Exit Slip	Collect and analyze all of the student work and sort them into three (or as many as appropriate) piles based on the strategies or types of errors used. In the next session, the teacher will work with small groups to help move them to their next step.

Does this work? Why or Why not?

$$\begin{array}{r} 40 \\ -28 \\ \hline 28 \end{array}$$

Name_____

73 − 68 =   **51 − 22 =**

45 − 17 =   **81 − 76 =**

  34   **27**

−18   **−16**

  51   **65**

−27   **−46**

Excerpt from the classroom:

Even though the problem was presented differently, it was still highlighting the same error as in the previous lesson. After yesterday, most all the students realized that the strategy was incorrect and they were able to explain why, mostly using the "nick-name" stratey, but some used subtracting tens and ones. It was clear today which students still carried the misconception of 0-8 = 8. In today's discussion for the few hold outs, we acted it out like packs of cookies (10 to a pack) and I am holding a plate for the individual cookies. I say that on the table next to me there are 4 packs of cookies and I am holding a plate, but there is nothing on it. To take 8 cookies off the plate, the student realizes that he will have to open the pack and empty it onto my plate. There is only one student who is left who does not seem to understand the connection and he is now going to be a target for an intensive small group or 1-1 instruction.

When the practice sheets were analyzed, there were essentially three groups. One group understood the idea of regrouping, but may have showed it in different ways. This group was ready to clean up their recording system to a more efficient and cleaner one. The second group comprised of students that perhaps had some errors and some correct answers. They were trying to use strategies, but maybe in erroneous ways. They just needed a little boost to make sure they were holding on to the meaning of the numbers. Most of the students in the third group were doing the typical errors children make when subtracting, such as just taking the absolute difference between the two numbers. These small groups were formed for the next lesson.

Class Notes:

Practicing Subtraction Procedures

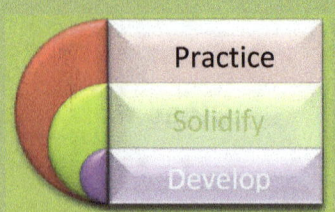

Day: 28-29

Objective: In the context of working in small groups, students will be able to move to their next developmental level in regards to how they are dealing with subtraction.

Materials: Math journals and pencil.

Small group time

Teacher will work with students in small groups. What the other students are doing is up to the individual teachers. Suggestions might be to work on math facts, review previous skills, or engage in a problem solving activity with a partner or triad.

At this point, it is most likely that there will be 3 groups of students, but this might vary from class to class.

<u>Students who understand regrouping or renaming as a way to subtract, but just need to clean up their recording system (practice phase):</u> With this group, take some examples from their practice sheet the day before and write a few of their strategies, one at a time, to the same problem, and make sure that all can articulate what is happening in the problem. Pick the strategy that most mirrors the standard algorithm and model a cleaner way to record, eliminating arrows and unnecessary markings, since now we can all assume that they understand what they are doing. Model recording both horizontal and vertical problems. Have students practice independently with 2 horizontal and 2 vertical problems. This process should be rather quick, taking no more than 15 minutes. Consider giving these students a handout that introduced three digit addition and subtraction to see how they transfer concepts before it is officially introduced in the classroom. Encourage them to work on it in pairs.

<u>Students who are using strategies, perhaps erroneously, but are showing understanding of tens and ones and are trying to preserve place value (solidify / practice phase):</u> The work with this group will depend on what they are trying to do. Just let them know that you are going to help streamline their strategies and focus on the renaming (a.k.a., "nicknaming") strategy. Write a problem on the board, have students tell you what you should do next for each step. Depending on the nature of their errors, this group should also take about 15-20 minutes. It is most likely that in this group it will be necessary to individually conference

with students if they need extra support. If necessary, consider pulling those from this group who are still struggling for an extra small group time the following day, or allowing a student from the first group to partner up with the student to support him/her while doing practice problems.

<u>Students who are making typical errors that demonstrate they are not preserving place value when working with the numbers (solidify phase)</u>
It is likely that there will be students in your classroom who are still or need to make the appropriate connections between what they do with the base ten blocks and what the numbers mean. The work with this group will be significantly longer and it will be essential to go back and work with the base ten blocks. Begin by having one set of blocks for all to see. Pose a problem on the board and ask them what to do with the blocks, then do what they say, or have one student do it. Then ask how they can record that in the numbers. If it matches, write what they say. A lot of attention needs to be on what they are building and the connections to the abstract numbers. All the students should know by now how to build the problem, but they need more time making the direct connections to how to record it. After a few rounds, pose a problem and have them build and record on their own, to see who are making the connections and who are not. Modify the group and instruction as needed.

Afterwards Continue small groups as needed. The students still in the solidify phase will need a few more days. Find ways to meet with them as you begin to move on to other concepts.

Excerpt from the classroom:

The following problems are examples of two students who showed understanding of tens and ones and renaming when subtracting, but perhaps had an inefficient recording strategy.

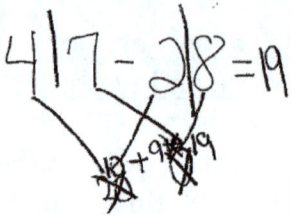

After a 10 minute small group, the same students were able to solve similar problems in the following fashion:

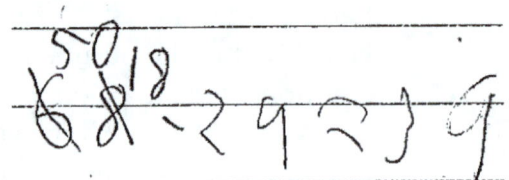

The students in the middle group had, in the course if the long break, forgotten some of the rationale of what they were trying to do, but the caught up very quickly and in the same amount of time were able to achieve efficient strategies. Below is an example of a student's before and after strategy. At first, it appears that she was trying to make a ten with the eight, a strategy from addition, but then added the two and four together. Next is her strategy after just 15 minutes of clearing up her misconception, and she was able to maintain this strategy from then on out.

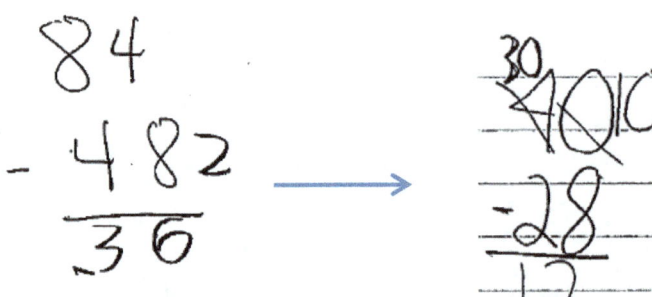

The students in the third group, who were still making typical errors regarding subtraction, were given intensive small group instruction time for two days to build and make specific connections between each step and recording it numerically. By the end, all but two were able to accurately subtract with an efficient recording strategy.

While doing continued intensive instruction with the more struggling group, the rest of the students were doing practice problems. Students in the first two groups were given the challenge of trying to figure out a couple of three digit addition and subtraction problems, as a pre-assessment for that unit. Other students were given more two digit problems to practice to build automaticity.

Class Notes:

Assess & Celebrate

Day: 30

Class Notes:

Works Cited

Bahr, D. L. (in progress). *An integrated model of teacher change: A theoretical framework for conceptualizing teacher learning in an era of mathematical reform.*

Bahr, D. & L.A. de Garcia. (2010). *Elementary Mathematics is Anything but Elementary: Content and methods from a developmental perspective.* Belmont, CA: Cengage.

Carpenter, T.P., Fennema, E., Franke, M. L., Levi, L. & Empson, S. (1999). *Children's mathematics: Cognitively guided instruction.* Portsmouth, NH: Heinemann.

Common Core State Standards: Math (http://www.corestandards.org/Math/)

Dekker, T. (2007). The Dutch experience, threat or treat? In S. Close, D. Corcoran, and T. Dooley (Eds.) Proceedings of: *Second National Conference on Research in Mathematics Education MEI 2.* Dublin, Ireland: St. Patrick's College.

Kilpatrick, J., Swafford, J., & Findell, B., (eds.). (2001). *Adding it up: Helping children learn mathematics.* Washington, DC: National Academy Press.

Lesh, R. A., Post, T. R.. & Behr, M. J. (1987). Representations and translations among representations in mathematics learning and problem solving. In C. Janvier (Ed.), *Problems of representation in the teaching and learning of mathematics,* (pp. 33–40).

National Council of Teachers of Mathematics. (1991). *Professional standards for teaching mathematics.* Reston, VA: National Council of Teachers of Mathematics.

Schifter, D., and Fosnot, C.T. (1993). Reconstructing mathematics education. New York: Teachers College Press.

Solution Tree (2010). *Making math accessible to students with special needs.* Bloomington, IA: Solution Tree Press.

Woodward, J. (2006). Making reform-based mathematics work for academically low-achieving middle school students. In M. Montague & A. K. Jitendra (Eds.) *Teaching mathematics to middle school students with learning difficulties* (pp. 29-50). New York City, NY: Guilford Press.